T0094370

Springer Optimization and Its Applications

VOLUME 71

Managing Editor
Panos M. Pardalos (University of Florida)

Editor–Combinatorial Optimization
Ding-Zhu Du (University of Texas at Dallas)

Advisory Board
J. Birge (University of Chicago)
C.A. Floudas (Princeton University)
F. Giannessi (University of Pisa)
H.D. Sherali (Virginia Polytechnic and State University)
T. Terlaky (Lehigh University)
Y. Ye (Stanford University)

Aims and Scope
Optimization has been expanding in all directions at an astonishing rate during the last few decades. New algorithmic and theoretical techniques have been developed, the diffusion into other disciplines has proceeded at a rapid pace, and our knowledge of all aspects of the field has grown even more profound. At the same time, one of the most striking trends in optimization is the constantly increasing emphasis on the interdisciplinary nature of the field. Optimization has been a basic tool in all areas of applied mathematics, engineering, medicine, economics, and other sciences.

The series *Springer Optimization and Its Applications* publishes undergraduate and graduate textbooks, monographs and state-of-the-art expository work that focus on algorithms for solving optimization problems and also study applications involving such problems. Some of the topics covered include nonlinear optimization (convex and nonconvex), network flow problems, stochastic optimization, optimal control, discrete optimization, multiobjective programming, description of software packages, approximation techniques and heuristic approaches.

For further volumes:
http://www.springer.com/series/7393

Nicholas J. Daras

Editor

Applications of Mathematics and Informatics in Military Science

 Springer

Editor
Nicholas J. Daras
Department of Military Science
Hellenic Army Academy
Vari Attikis
Greece

ISSN 1931-6828
ISBN 978-1-4614-4108-3   ISBN 978-1-4614-4109-0 (eBook)
DOI 10.1007/978-1-4614-4109-0
Springer New York Heidelberg Dordrecht London

Library of Congress Control Number: 2012942844

Mathematics Subject Classification (2010): 11T71, 14G50, 34L25, 34A30, 34A34, 37L30, 37N40, 41A05, 41A20, 41A21, 42A16, 51M04, 51N20, 60G25, 60J10, 60K15, 65C20, 65C50, 65D18, 65S05, 65T60, 68P25, 68P30, 68U20, 74J20, 78A46, 81U40, 90B10, 90B50, 90C11, 90C29, 90C90, 94A05, 94A15, 94A60, 97R30, 97R60, 97U70

Printed on acid-free paper

Springer is part of Springer Science+Business Media (www.springer.com)

Foreword

Today, many practitioners and students of war approach it as a discipline founded on purely scientific principles. The mathematics involved in decision making are playing an increasingly important role in military tactics and applications.

This book is the outgrowth of a conference held in April 2011 at the Hellenic Army Academy and aims to examine the application of mathematics and informatics military science. Essentially, it is an attempt to synthesize the concepts of military science, together with the concepts and techniques of applied operations research, signal processing, scattering, scientific computing and applications, simulation, satellite remote sensing, coding, and statistical modeling. The work will prove useful as a textbook and reference for these subjects; it could also be a basis for further study and research.

The key features of the book and their corresponding benefits are:

1. The leveraging of analysis in support of current operations in order to improve analytic support to the war fighter and to improve analytical capability for warfare
2. To working groups meeting in composite sessions to address a wider spectrum of topics
3. To the development of courses of action or methodologies to reconcile issues identified
4. To enable cooperation between various scientific and military communities by bringing together a range of interdisciplinary objects across a breadth of military areas

The book's 15 chapters are independent of each other and cover so many scientific aspects of military science that only a well-read military analyst would be conversant with them all. The suggestion is to approach the book as a whole. We hope that the book will be especially helpful to students and new entrants to the field, as well as to people whose knowledge is already established.

NATO C3 Agency   Georges D'Hollander

Preface

This work is an outgrowth of a conference held in April 2011 at the Hellenic Army Academy and brings together a wide variety of mathematical methods with application to defense and security and discusses directions and pursuits of scientists that pertain to the military. Also studied is the theoretical background required for methods, algorithms, and techniques used in military applications as well as the direction of theoretical results in these applications. Open problems and future areas of focus are also highlighted.

Topics covered include applied operations research and military applications, signal processing, scattering, scientific computing and applications, simulation and combat models, satellite remote sensing, coding, statistical modeling, and applications. Analysis, assessment, and data management are core competencies for operations research analysts. At the two day colloquium "Applications of Mathematics and Informatics of Military Education", held at the Hellenic Army Academy April 11–12, 2011, we addressed these issues and developed production recommendations for improving our analysis, assessment, and data management ability.

One current need is to review the assessment framework in order to catalog best practices. This process will also help to better identify the data needed for assessments. Armed with a clearer assessment framework, the operations research community could then work with the operational community to improve the assessment and data management processes.

One of the largest issues we have in trying to develop assessment decision support using multi-criteria analysis methods is that the objectives are not always clearly articulated. From these objectives should flow what we do for assessments, but often the reverse is true—we assess what we can and then provide that to senior commanders in lieu of what they might need.

Additional aspects needed by the broader community are software, statistical and computing tools used to support the process and methods for displaying the information for a broad number of metrics that provide meaning and insight into objective accomplishment.

We also need to better educate the operations community so that the leaders have realistic expectations of what can be produced from warfare analysis and assessment. Commanders need to understand that much of the warfare assessment process relies on qualitative data as opposed to the quantitative data often used in assessing conventional campaigns. One potential product of the "Applications of Mathematics and Informatics of Military Education", held at the Hellenic Army Academy April 11–12, 2011 would be a presentation that clearly outlines the current state of assessment capabilities in a warfare environment. The information derived from the presentation could then be shared with Division and Corps commanders to help them better understand the assessment process.

Finally, the assessment process involves analysis across multiple lines of operation (satellite remote sensing, security and information operations). One beneficial product would be a community wide campaign plan for combat modeling. The plan would be designed to answer key commander questions.

Vari Attikis, Greece   Nicholas J. Daras

Contents

Part I
General

Chapter 1
The Significance of Research and Development for National Defence and Its Relation with the Military University Institutions

Nikolaos Uzunoglu

Abstract In this article, the difficulties and obstacles being faced in the effort to implement technology-oriented research and development in the area of national defence applications are presented. In addition, the importance of investing human research resources for the benefit of National Defence is emphasized based on the author's 35 years of experience in Greece. As a last point, the significance of Military Universities offering graduate programs is stated, which could develop into a strong factor to support National Defence and also stimulate local industry.

Keywords Research and Development

Mathematics Subject Classification (2010): 00A99, 97B10, 97M99

1.1 A Short Historical Overview

The history of mankind is full of conflicts and wars in which technological initiatives played a decisive role. To mention a few examples:

- In the ancient era: Starting from the Trojan war up to the illustrative examples of Archimedes' defence technologies and the Athenian fast moving ships—the triremes.
- In the Twentieth Century, the two World Wars: The proper use of technology changed the course of the war. The invention of the microwave radar, a result of the research efforts of a small research group at the University of Birmingham (magnetron), and computational principles were used to attack

N. Uzunoglu (✉)
National Technical University of Athens, School of Electrical and Computer Engineering, Spetson 1A, GR 15122, Maroussi, Greece
e-mail: nuzu@cc.ece.ntua.gr

N.J. Daras (ed.), *Applications of Mathematics and Informatics in Military Science*, Springer Optimization and Its Applications 71, DOI 10.1007/978-1-4614-4109-0_1, © Springer Science+Business Media New York 2012

difficult-to-break codes. In contrast, despite the advanced nature of the missile technology developed by Germany, it did not have any effect on the course of the war, not being used at the proper time.

– During the Cold War: The competition in the field of technology and the inability of the Soviet side to compete with the western alliance in the fields of computers and communications (because of wrong strategic selections) determined the outcome of the cold war.

– During the last 20 years: The developments in non-military technologies have overtaken the military technologies, opening up the possibility of developing intelligent and clever systems based on commercial components provided a country possesses the necessary experienced engineers, human resources, and laboratory infrastructure.

Analysis of the present defence market clearly indicates that technologically strong state entities are spending much political effort to convince the technologically weak states to allocate substantial funds to defence projects, which are highly dependent on technology and its role in international relations. This observation is highly linked with the debt burden of technologically weak states.

1.2 The Significance of National Technology for National Defence

In a sovereign country, the necessary conditions to develop national defense technologies are as follows:

- The political determination of the country's leadership to pursue a truly sovereign policy.
- The creation of a technological research and development capacity by investing in human resources as well as technological infrastructure.
- The establishment of significant potential for industrial production. In this framework, the main issues to be addressed are the following:
- Deciding the country's production policy: whether to invest in creativity and innovation or in a commercial and service economy approach.
- Choosing feasible objectives and implementing a long-term program in upgrading the national defence infrastructure.
- The collaboration of defence agencies with research institutions and national industry with well-defined rules.
- Orienting the entire system of defence armament programs to support the nationalization of production so that it reaches a significant level, ensuring saving of important national resources that will be forwarded to other vital social sectors such as education and health.
- Initiatives by the national research capital that will support the effort of upgrading and exploiting technology for the benefit of national defence.
- Focusing the domestic industry, public or private, on "real production" and, at the same time, abandoning the syndrome of "commercial representative".

1.3   Specific Observations of the Last 35 years in Greece

Many obstacles are being faced in Greece over the last 35 years related to the national defence industry:

- The legal background: The whole system of legislation has been oriented to providing open floor solely to technology users of the armed forces and this has led to powerful lobby groups placing insurmountable obstacles against the efforts to develop prototype technological advancements in the field of defence systems.
- The Law on Inventions and Patents concerning National Defence: Despite the fact that a Law going as far back as 1964 foresaw the existence of a Committee under the auspices of the Ministry of Defence to review the patents to be submitted to receive approval and in case a patent has significant value for national defence, to be classified and treated as such, this procedure has been abandoned at least in the last 15 years.
- Research Institutes: Presently there are several research units and establishments that have the capability to carry out original work in the field of research and development related to national defence projects. This potential has not been exploited in full.
- The History of Matching Programs: It is a well-known fact that the funds allocated to armament programs during the last 30 years have not been used to stimulate national production programs; rather, they were used to increase the profit of representatives of foreign companies without any concrete results. This is contrary, what other countries have achieved by utilizing the funds in corresponding cases by developing production or research and development facilities.
- Assessment of the obtained results of the Research Efforts: Despite the difficulties, there have been several serious efforts in the development of prototypes based on novel ideas incorporating original ideas. Of course, an independent assessment of these efforts is needed to arrive to a concrete conclusion.
- Research Potential of Human Resources: It is important to state that, in terms of human resources, Greece has a unique capital of scientists, taking into account the human potential inside and outside the Country. However, this potential should be utilized with proper management.

1.4   The Need for Extending Military Universities into R & D Through Postgraduate Studies

The political and military leadership of the National Defence Ministry should set the objectives for the exploitation of the human capital of the military Universities for the support and development of the research needs of the armed forces. The realization of this objective could be served best by the present military universities corresponding to the land, sea and air armed forces. Specifically, this could be achieved by the establishment of postgraduate programs corresponding to sectors that are vital for national defence. Following the National Defence Ministry's definition of priority sectors and fields that needed to be developed for the benefit of national defence, an expert committee should be established to propose, in the short term, specific postgraduate programs. These postgraduate programs will be supervised by a special committee that will be formed and authorized by special government law. The basic guidelines for these postgraduate programs should be as follows:

- Intensive and focused education
- Diploma thesis: Conveying of significant and vital operational problems.
- PhD studies in cooperation with University Laboratories and Research Institutes.
- The potential of officers with postgraduate or doctoral degrees should be utilized by appointing them as instructors and supervisors of projects.

1.5   Conclusions

In this article, the problems being faced in Greece in the field of R & D in national defence are analyzed and then a specific proposal is submitted on the extension of the three military universities to postgraduate studies by incorporating research and development activities to serve the urgent needs of national defence. In this effort, the role of the military universities will be crucial after the extension of their activities to R & D. The design of the postgraduate framework should (a) minimize the financial burden, (b) exploit the nations scientific capital, and (c) prevent external factors the influence of that could jeopardize the whole effort.

Part II
Applied Operational Research
and Military Applications

Chapter 2
Selected Topics in Critical Element Detection

Jose L. Walteros and Panos M. Pardalos

Abstract In this paper we consider the problem of detecting critical elements in networks. The objective of these problems is to identify a subset of elements (i.e., nodes, arcs, paths, cliques, etc.) whose deletion minimizes a given connectivity measure over the resulting network. This paper surveys some of the recent advances for solving these kinds of problems including heuristic, mathematical programming, approximated algorithms, and dynamic programming approaches.

Keywords Critical element detection • Critical node problem • Critical clique detection

Mathematics Subject Classification (2010): 90-02

2.1 Introduction

In network analysis, the problem of detecting subsets of elements important to the connectivity of a network (i.e., critical elements) has become a fundamental task over the last few years. Identifying the nodes, arcs, paths, clusters, cliques, etc., that are responsible for network cohesion can be crucial for studying many fundamental properties of a network. Depending on the context, finding these elements can help to analyze structural characteristics such as attack tolerance, robustness, and vulnerability. Furthermore, it can also help for classifying members based on their centrality, prestige, and reputation; and to determine dominant clusters and partitions.

J.L. Walteros (✉) • P.M. Pardalos
Department of Industrial ans Systems Engineering, Center for Applied Optimization, University
of Florida, 303 Weil Hall, Gainesville, FL, USA
e-mail: jwalteros@ufl.edu; pardalos@ise.ufl.edu

N.J. Daras (ed.), *Applications of Mathematics and Informatics in Military Science*,
Springer Optimization and Its Applications 71, DOI 10.1007/978-1-4614-4109-0_2,
© Springer Science+Business Media New York 2012

From the point of view of robustness and vulnerability analysis, evaluating how well a network will perform under certain disruptive events plays a vital role in the design and operation of such a network. To detect vulnerability issues, it is of particular importance to analyze how well connected a network will remain after a disruptive event takes place, destroying or impairing a set of its elements. The main strategy is to identify the set of critical elements that must be protected or reinforced in order to mitigate the negative impact, the absence of such elements may produce in the network. Applications of this kind are typically found in homeland security [15, 17], evacuation planning [21], immunization strategies [28], energy [24], and transportation [19].

From the member-classification perspective, identifying members with a high reputation and influential power within a social network could be of great importance when designing a marketing strategy. Positioning a product, spreading a rumor, or developing a campaign against drugs and alcohol abuse may have a great impact over society if the strategy is properly targeted among the most influential and recognized members of a community. The recent emergence of social networks such as Facebook, Twitter, LinkedIn, etc. provides countless applications for problems of critical-element detection.

Furthermore, determining dominant cliques or clusters over different industries and markets via critical clique detection may be crucial for analyzing market share concentration, debt concentration, spotting possible collusive actions and may even help to prevent future economic crisis.

This paper surveys some of the recent advances for solving these kinds of problems including heuristics, mathematical programming, approximated algorithms, and dynamic programming approaches. We provide a brief description of the mathematical models and formulations used in the referenced papers. Proofs, computational experiments, and additional discussions are in general omitted, but sources are fully referenced in each case. To avoid discrepancies with the sources, we present the mathematical formulations using the original notation as found in each paper. There are few cases though, where we are forced to slightly modify some elements to ensure the consistency in the style. However, we try our best effort to point out these differences to avoid possible confusion.

In general, critical element detection problems lie in the boundaries of different research areas including network interdiction [18, 30], network design [13], graph clustering and partitioning [14, 26], among others. We provide the reader with additional references to some related problems in these areas; however, we do not include detailed descriptions. While we have tried not to omit any of the recent publications, we have no claim to completeness.

We organize this paper as follows. In Sect. 2.2 we give a general description of the critical element detection problem and summarize some of the different connectivity measures that are generally used in the literature. Section 2.3 focuses on the critical node problem (CNP). We describe three mathematical formulations and present some additional methodologies used to solve this problem. In Sect. 2.4 we describe the literature regarding the critical arc detection problem. In Sect. 2.5, we introduce the critical clique detection problem (CCP). We provide two mathematical formulations designed for using two different connectivity measures. Finally, in Sect. 2.6 we provide conclusions and further directions for subsequent projects.

2.2 Critical Element Detection Problems

Let $\mathcal{G} = (\mathcal{V}, \mathcal{E})$ be a graph, where $\mathcal{V}$ is the set of nodes and $\mathcal{E}$ is the set of arcs. Let $\mathcal{T}$ be the set of elements to be analyzed. Let b_t be a cost associated with the deletion of element $t \in \mathcal{T}$ and let B be a deletion budget. In general, most critical element detection problems aim to identify a subset of elements in $\mathcal{T}$ (i.e., the critical elements) whose removal minimizes the connectivity of the residual network, while satisfying the budget constraint.

Depending on the context, $\mathcal{G}$ may vary in its composition. It can be a tree, a planar graph, a series–parallel graph, a forest, or a more complex graph. It can be defined as directed or undirected, and the degree of the nodes may follow different distributions (uniform, power law, etc.). Moreover, $\mathcal{T}$ may represent a set of different kinds of elements. It can simply be defined as the set of nodes or the set of arcs, or can be comprised by more complex substructures such as paths, cliques, or clusters. In the contexts of the critical node and the critical arc detection, we have that $\mathcal{T} = \mathcal{V}$ and $\mathcal{T} = \mathcal{E}$, and costs b_t are node and arc deletion costs b_i for $i \in V$ and b_e for $e \in E$, respectively. For more complex cases, such as for the CCP, set $\mathcal{T}$ is composed by all the possible cliques in $\mathcal{G}$ and the deletion cost b_t can be defined as the sum of the deletion cost of all the nodes and arcs of clique $t \in \mathcal{T}$. Additionally, in some papers the deletion cost b_t is assumed to be one for all the elements, therefore B is defined as the upper bound of the number of elements that can be deleted.

In addition to the structure of the network and the definition of the critical elements, a crucial factor that should be taken into account is the measure that is used to analyze the connectivity of the residual network. There are several ways to quantify how well connected a network is, and depending on the measure that is chosen, the complexity of the problem, the solution approach, and the optimal results may vary dramatically. Even though the principal objective is to find the elements whose absence disconnects the network the most, depending on which measure is used to calculate the connectivity of the residual network we may obtain different optimal solutions. As we will point out in Sect. 2.3, a simple modification in the structure of the costs may lead to a significantly more difficult problem. Moreover, the selection of appropriated solution technique also depends on the connectivity measure. Some measures are easier to formulate with some techniques than with others. We now summarize some of the measures that are commonly used.

2.2.1 Connectivity Measures

Several measures have been used to assess the level of disconnection of the residual network. These measures can be categorized into two classes depending on the context of the problem that is being solved. The measures from the first class are mainly associated with network flow problems, in particular shortest path problems, maximum flow problems, or minimum cost flow problems. The logic behind these measures is that a network gets disconnected when it starts losing its ability to send flow between a predefined set of node pairs, or simply when traversing the network becomes too expensive (see [7, 16, 30]). For these cases, the critical elements are the ones whose deletion results in the maximum increase of the shortest paths or, consequently, the maximum decrease of the flow capacity between the predefined node pairs. These kinds of measures are commonly used in the context of network interdiction (see [6, 18, 20, 21, 31]), and are generally designed to tackle arc interdiction problems (detecting critical arcs).

On the other hand, the measures of the second class are mostly associated with topological characteristics of the network. Among this class, the most common measures are: the total number of pairwise connections (i.e., the total number of node pairs that are connected in the network by at least one path) [3, 8], the total weighted pairwise connectivity (i.e., a weighted sum of the pairwise connections) [3, 8], the size of the largest connected component (i.e., the number of nodes that belong to the largest maximal connected subgraph of $\mathcal{G}$) [23, 25], and the total number of connected components [2, 25].

The first two measures of this class are mainly explicit evaluations of how reachable the nodes of the network are in the absence of the critical elements. These measures are generally used when there is no need to account for capacities or costs associated with the paths between the nodes. For example, these can be used when analyzing the vulnerability of a city's infrastructure during an evacuation event to identify areas of the city that can be totally disconnected from predefined safe zones.

The measure that accounts for the size of the largest component can be used to achieve a relatively more homogeneous disconnection. When minimizing the size of the largest component, despite a possible sacrifice in the total number of pairwise connections, we can avoid having large concentrations of connections in the residual network. A possible application for this measure could be to identify the key members of a criminal organization that have to be captured to maximize the segregation of the remaining members.

Additional topological measures that can be incorporated in critical element detection problems may involve the diameter of the residual components (i.e., the shortest path between the two most remote nodes in the network), the degree of the remaining nodes (i.e., the number of incident arcs to a node), the number of paths between every pair of nodes, and the number of common neighbors every pair of the remaining nodes, among others. For additional ways of measuring vulnerability and connectivity see [5, 15].

In general, the selection of the adequate measure is fundamental for making the right analysis. Despite the fact that all of these account for a connection level of the given network, using one over the other may lead to a completely different set of critical elements. Figure 2.1 provides an example of the different optimal solutions that are found depending on the measure that is chosen. For this example we assume that the goal is to identify the most critical node among the network. Note that if the

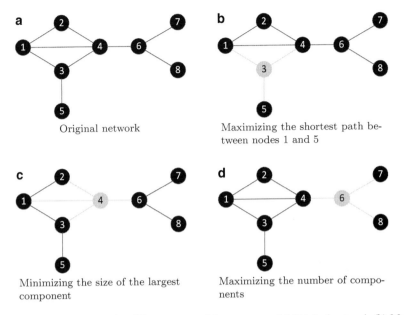

Fig. 2.1 Optimal solutions for different connectivity measures. (**a**) Original network (**b**) Maximizing the shortest path between nodes 1 and 5 (**c**) Minimizing the size of the largest component (**d**) Maximizing the number of components

connectivity measure is set to be the length of the shortest path between nodes one and five, the critical node is node three. On the other hand, if the measure is the size of the largest component, the critical node is node four. And finally, if the measure is the total number of components, the critical node is node 6.

2.3   Critical Node Detection Problems

Among all of the critical element detection problems, the ones of detecting critical nodes and critical arcs are the ones that have attracted significantly more attention. In this section we consider the problem of detecting critical nodes on graphs, often referred as the CNP.

From the complexity point of view, the decision version of the CNP was proven to be $\mathcal{NP}$-complete on general networks in [3, 9, 23], by reductions from either the vertex cover problem or the independent set problem [11]. These complexity proofs were developed for the total number of pairwise connections and the total weighted pairwise connectivity measures, but they can be generalized for other as well.

An additional complexity analysis of the CNP for other network topologies such as trees and series–parallel graphs can be found in [8, 25]. Di Summa et al. [8] proved that the CNP is also $\mathcal{NP}$-complete on trees for the total weighted pairwise connectivity measures using a reduction from the multicut in trees problem [12]. They also showed that when the pairwise connection costs are set to be one (i.e., when the connectivity measure is replaced with the sum of the total number of pairwise connections), the problem can be solved in polynomial time using a dynamic programming approach (see Sect. 2.3.4). Moreover, Shen and Smith [25] proved that the CNP is polynomially solvable in trees and series–parallel graphs for the cases when the deletion costs of the nodes are set to be one and the objective is either minimizing the size of the largest component or maximizing the number of residual components. They also propose a dynamic programming scheme for solving the CNP in these cases (see Sect. 2.3.4).

We now discuss the literature regarding the existing methodologies for solving the CNP. These include mathematical programming, approximated algorithms, heuristics, and dynamic programming approaches. We first introduce some mathematical formulations and then give a short description of other approaches.

2.3.1 Mathematical Formulations

When studying combinatorial problems, using a mathematical formulation is in general a natural starting point. Despite the inherent difficulty of these problems, techniques such as branch-and-bound and branch-and-cut are proven to be very efficient approaches to obtain solutions for instances of manageable size. Recent endeavors using mathematical programming techniques can be found in [3,6,16,23].

The mathematical formulation introduced in [3] is designed to tackle the CNP for the case when the total number of pairwise connections is used as the connectivity measure and when the deletion costs are set equal to one. Note, however, that this formulation can be easily adapted to use the weighted pairwise connectivity or to include deletion costs different than one. For this formulation, the authors define a binary variable u_{ij} for every pair of nodes $i, j \in V$ that takes the value of one if nodes i and j belong to the same component and zero otherwise. In addition, they introduce a binary variable v_i for every node $i \in V$ that takes the value of one if node i is deleted in the optimal solution and zero otherwise. The mathematical formulation is as follows:

$$\min \sum_{i,j \in V} u_{ij} \tag{2.1}$$

$$\text{s.t. } u_{ij} + v_i + v_j \geq 1 \qquad\qquad (i,j) \in \mathcal{E} \tag{2.2}$$

$$u_{ij} + u_{jl} - u_{il} \leq 1 \qquad\qquad i,j,l \in V \tag{2.3}$$

$$u_{ij} - u_{jl} + u_{il} \leq 1 \qquad\qquad i,j,l \in V \tag{2.4}$$

$$-u_{ij} + u_{jl} + u_{il} \leq 1 \qquad\qquad i,j,l \in V \tag{2.5}$$

$$\sum_{i \in V} v_i \leq B \tag{2.6}$$

$$v_i \in \{0,1\} \qquad\qquad i \in V \tag{2.7}$$

$$u_{ij} \in \{0,1\} \qquad\qquad i,j \in V \tag{2.8}$$

where the objective function (2.1) minimizes the sum of pairwise connections. Constraint (2.2) ensures that if there is an arc between two nodes and none of them is deleted, they have to be in the same component. Constraints (2.3)–(2.5) are known as triangle inequalities and ensure the transitive relationship between the connections in the graph (i.e., if node i is connected to node j and node j is connected to node l, then node i must also be connected with node l). Constraint (2.6) sets the upper bound on the number of critical nodes, and constraints (2.7) and (2.8) define the domain of the variables used.

An alternative formulation for the CNP was presented in [16]. In this paper, the author's aim is to analyze what is the maximum (and the minimum) possible disruption that can occur in the network when a given number p of nodes are removed. The idea behind this approach is that, by calculating the maximum and the minimum damage that can be inflicted to the network, one can obtain a broader picture of the possible damages that may occur during a real disruptive event. To account for the disruption level, the authors define a parameter a_e for each arc $e \in \mathcal{E}$, which measure the demand volume that is present in arc e in the case that the arc remains in the network after the removal of the critical nodes. For this formulation, let $\mathcal{E}(i)$ be the set of arcs that are incident to node i and $r_i = |\mathcal{E}(i)|$ be its size. Consequently, let $\mathcal{V}(e)$ be the set of end points $(i,j) \in V$ of arc e. Let x_i be a binary variable that takes the value of one if node i is deleted and zero otherwise. Finally, let y_e be binary variable that takes the value of one if arc e remains in the network and zero otherwise. The mathematical formulation now follows. We slightly modify the notation of this formulation to be consistent with the style of this paper.

$$\max(\min) \sum_{e \in E} a_e y_e \tag{2.9}$$

$$\text{s.t.} \sum_{i \in V} x_i = p \tag{2.10}$$

$$\sum_{e \in \mathcal{E}(i)} y_e \leq r_i(1 - x_i) \qquad \forall i \in V \tag{2.11}$$

$$y_e \geq 1 - x_i - x_j \qquad e \in \mathcal{E}, (i,j) \in \mathcal{V}(e) \tag{2.12}$$

$$x_i \in \{0,1\} \qquad\qquad i \in V \tag{2.13}$$

$$y_e \in \{0,1\} \qquad\qquad e \in \mathcal{E} \tag{2.14}$$

where the objective function (2.9) either maximizes or minimizes the sum of arc demands that are not affected by the node disruption. Constraint (2.11) sets the number of nodes that will be removed. Constraints (2.11) ensure that if node i is deleted, the arcs incident to it are not present in the final network. Constraints (2.12) enforce that an arc is present in the final network if none of its end points is removed. Finally, constraints (2.13) and (2.14) define the domain of the variables used. Alternative formulations regarding the CNP in this context can be found in [6].

A different mathematical formulation was introduced in [23]. In this work, the formulation is designed to solve a simple variation of the CNP where the objective is to minimize the number of nodes deleted, while ensuring that the size of the residual components is less or equal than a given parameter c. In this formulation, variable x_{ij} is a binary variable that takes the value of one if nodes i and j belong to the same component and zero otherwise, and y_i is a binary variable that takes the value of one if node i is not deleted in the optimal solution and zero otherwise (note that in this formulation variables y_i are defined in the opposite way as variables v_i are defined in [3]). The mathematical formulation follows:

$$\max \sum_{i \in \mathcal{V}} y_i \tag{2.15}$$

$$\text{s.t. } u_{ij} + u_{jl} - u_{il} \leq 1 \qquad\qquad i, j, l \in \mathcal{V} \tag{2.16}$$

$$u_{ij} - x_{jl} + x_{il} \leq 1 \qquad\qquad i, j, l \in \mathcal{V} \tag{2.17}$$

$$-x_{ij} + x_{jl} + x_{il} \leq 1 \qquad\qquad i, j, l \in \mathcal{V} \tag{2.18}$$

$$\sum_{i \in \mathcal{V} \setminus \{i\}} x_{ij} \leq c - 1 \qquad\qquad \forall i \in \mathcal{V} \tag{2.19}$$

$$y_i + y_j - x_{ij} \leq 1 \qquad\qquad (i, j) \in \mathcal{E} \tag{2.20}$$

$$y_i \in \{0, 1\} \qquad\qquad i \in \mathcal{V} \tag{2.21}$$

$$x_{ij} \in \{0, 1\} \qquad\qquad i, j \in \mathcal{V} \tag{2.22}$$

where the objective function (2.15) maximizes the sum of nodes that are not deleted. Constraints (2.16)–(2.18) are the triangle inequalities; constraints (2.19) ensure that each of the residual components is comprised by no more than c nodes; constraints (2.20) ensure that if there is an arc between two nodes and none of them is deleted, they have to be in the same component. And finally constraints (2.21) and (2.22) define the domain of the variables used.

In this paper, the authors also provide a detailed polyhedral analysis of this formulation and discuss some valid inequalities that are inherited from some clique partitioning problems. We now discuss the heuristic approaches.

2.3.2 Heuristics Approaches

A simple heuristic approach regarding the CNP was used in [2]. Rather than detecting which are the set of critical nodes of the network, this work is aimed to

Fig. 2.2 Suboptimal solution found by greedy approaches vs the optimal solution

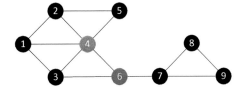

study the tolerance of complex networks with respect to strategic node deletions. The authors seek to analyze the resulting consequences over the network when nodes with a relative high importance are removed. In this case, the authors used the degree of the nodes (i.e., the number of arcs that are incident to the node) as a measure of importance, and then analyze the network cohesion after deleting the nodes with the largest degree. However, even though it seems natural that removing a node with a larger degree may cause a large disconnection in the network, it is easy to show that this is not necessarily the case. For example, observe the network described in Fig. 2.2. Note that removing the node with the largest degree (node 4) does not affect the connectivity of the residual network at all, whereas removing a node with a lesser degree (node 6) divides the network into two components.

An alternative approach to solve the CNP is to use a greedy algorithm. For the case of the CNP, note that finding the node whose remotion minimizes the connectivity of the network (i.e., solving the CNP when the node-deletion budget is one) can be done in polynomial for any connectivity measure by removing individually each node from the network, and then solving a breadth-first search, a shortest path problem, or a maximum flow problem (see [1]), depending on the measure, to identify which is the node deletion that gives a better disconnection. Thus, a simple greedy algorithm for solving a general case of the CNP is to sequentially eliminate the node that generates the largest disruption at each iteration. Unfortunately, greedy algorithms like this are known to perform poorly in practice. In order to improve the suboptimal solutions that are obtained when using greedy techniques, Borgatti [4] introduces a local search heuristic. In this algorithm, an initial collection of critical nodes S is selected either randomly or with a greedy algorithm. Then, the local search performs a swap between each pair of nodes (s,t) such that $s \in S$ and $t \in V \setminus S$. If the swap leads to an improvement, the swap is accepted and set S is updated; otherwise, the algorithm continues with the exploration. The algorithm stops when no further improvements are found.

A greedy heuristic that is based on identifying maximal independent sets was proposed in [3]. The intuition behind this approach is that the subgraph induced by a maximal independent set is empty. Therefore, the deletion of the nodes that are not in the independent set will result in an empty subgraph. Note that if the size of the maximal independent set that is found by the algorithm is greater than the number of nodes in the network minus the critical node budget, the optimal solution for the CNP is to select the nodes that are not in the independent set. However, if the size of the independent set is less than this value, one can greedily keep adding nodes which provide the best improvement on the objective value until reaching the upper bound.

To improve the solution of the proposed algorithm, a local search procedure similar to the one described above is also implemented and is embedded into a multi-start heuristic, that sequentially reproduces the same algorithm and returns the best overall solution.

2.3.3 Approximation Algorithms

Form the approximation algorithms perspective, a variation of the CNP problem is presented in [9]. In this work, the authors propose a reformulation for the CNP where the objective function is set to minimize the number of nodes that must be removed in order to achieve a certain degradation (disruption) in the connectivity that is measured using the total pairwise connectivity measure. The authors provide a proof of the $\mathcal{NP}$-completeness of this version of the CNP by a reduction from the vertex cover problem. The authors also prove that this problem cannot be approximated within a factor less than 1.36 of the optimal solution, when the degradation level is set to be 0. Finally, they propose a $O(\log n \log \log n)$ pseudo-approximation scheme.

2.3.4 Dynamic Programming

The use of dynamic programming has been studied in [8, 25]. In the work of Di Summa et al. [8], the authors prove that the general version of the CNP over trees, for the weighted pairwise connectivity measure, is still $\mathcal{NP}$-Hard. For the cases where the total pairwise connectivity measure is used, with and without node deletion costs, they propose two dynamic programming algorithms with complexities $O(n^7)$ and $O(n^3 B^2)$, respectively.

In the work of Shen and Smith [25], the authors introduce a polynomial-time dynamic programming scheme for solving the CNP over trees and series–parallel graphs, for two connectivity measure: the number of connected components and the size of the largest component. For the case of the CNP over trees, the overall time and space complexities of the proposed algorithms are, respectively, $O(n^3)$ and $O(n^2)$ for the number of connected components measure and $O(n^3 \log n)$ and $O(n^2)$ for the size of the largest component. The authors also show that their approach can be slightly modify to tackle (with the same time complexity) the variant of the CNP where a deletion cost is present. Additionally, the authors proved that the CNP variation, where a weight is associated with each node and the objective is to minimize the largest weight over the components is $\mathcal{NP}$-Hard in the ordinary sense. They show that this variation can be solved with a pseudopolynomial version of their scheme.

For the case of the series–parallel graphs, Shen and Smith present two dynamic programming algorithms for the total number of connected components and the size of the largest component measure, with a time and space complexity of $O(n^3 \log n)$ and $O(n^2)$, and $O(n^5 \log n)$ and $O(n^3)$, respectively.

2.4   Critical Arc Problems

We now discuss some of the approaches for detecting critical arcs in networks including enumeration algorithms, mathematical programming approaches, and approximated algorithms.

2.4.1   Enumeration Approaches

An early approach for solving the critical arc detection problem by path enumeration was proposed in [7]. In this work the authors use as connectivity measure the shortest path between two predefined nodes s and t, and define the set of critical arcs $\mathcal{L}_n^* \subseteq A(D)$ as a set of n arcs such that the shortest path between s and t in $\mathcal{G}(\mathcal{V}, A(D) \setminus \mathcal{L}_n^*)$ is greater or equal than the shortest path between s and t in $\mathcal{G}(\mathcal{V}, A(D) \setminus \mathcal{L}_n)$ for any other set of n arcs $\mathcal{L}_n \subseteq A(D)$.

The idea of the algorithm is as follows. Let $\mathcal{P}$ be the set of all possible paths between s and t and $c(p)$ be the cost of path $p \in \mathcal{P}$. Let $\mathcal{P}^k = \{p_1^k, p_2^k, \ldots, p_{n_k}^k\}$ be the set of all kth shortest paths between node s and node t. That is, the set of paths such that $c(p_i^k) = c(p_j^k)$ and $c(p_i^h) < c(p_i^k) < c(p_j^l)$ for $h < k < l$. First, enumerate the set of all first shortest paths $\mathcal{P}^1$ and find the set of arcs $\mathcal{Q}^1$ that belong to the intersection of the paths in $\mathcal{P}^1$ (i.e., $\mathcal{Q}^1 = \bigcap_{i=1}^{n_1} p_i^1$). If this intersection is empty, all the arcs in the paths of set $\mathcal{P}^1$ are considered critical. On the other hand, if $\mathcal{Q}^1$ is not empty, find the set of all second shortest paths $\mathcal{P}^2$ and obtain set $\mathcal{Q}^2 = \bigcap_{i=1}^{n_2} p_i^2 \cap \mathcal{Q}^1$. If $\mathcal{Q}^2$ is empty, the set of critical arcs is set $\mathcal{Q}^1$. If set $\mathcal{Q}^2$ is a singleton, $\mathcal{Q}^2$ is the set of critical arcs. And, if $\mathcal{Q}^2$ is not empty and not a singleton, repeat the process for $\mathcal{Q}^3, \ldots, \mathcal{Q}^k$, until finding an empty set or a singleton set.

2.4.2   Mathematical Formulations

In most of the extant literature, the problem of detecting critical arcs is modeled as a network interdiction problem (e.g., [18, 20, 30, 31]). A network interdiction problem can be seen as a single-stage static Stackelberg game between a *leader* and a *follower* (see [27]), where the leader attempts to interdict a set of network elements, in an effort to optimally restrict the ability of the follower to use the network. For example, the leader may try to increase the cost that the follower perceives while traversing the network or to decrease the network capacity for shipping commodities. In contrast, the follower objective is either to minimize the total cost of using the network or to maximize the amount of commodities shipped. From the point of view of the critical element detection, the critical elements are then the elements that were optimally interdicted by the leader. That is, the elements whose deletion resulted in the maximum increase of the shortest paths used by the follower, or in the maximum decrease of the flow capacity of the network.

An example of a network interdiction formulation where the objective is to maximize the shortest path between two given nodes s and t can be described as follows. Let c_e be the cost associated with using the arc $e \in \mathcal{E}$. Let set $FS(i)$ and $RS(i)$ be the forward- and reverse-stars of node i, respectively. Let y_e be a binary variable that takes the value of one if the follower uses arc e and zero otherwise. Let x_e be a binary variable that takes the value of one if the leader interdicts arc e and zero otherwise. The mathematical formulation follows:

$$\max \min \sum_{e \in \mathcal{E}} c_e y_e \tag{2.23}$$

$$\text{s.t.} \sum_{e \in FS(i)} y_e \sum_{e \in RS(i)} y_e = \begin{cases} 1 & i = s \\ 0 & i \in \mathcal{V} \setminus \{s,t\} \\ -1 & i = t \end{cases} \tag{2.24}$$

$$y_e \leq 1 - x_e \qquad\qquad e \in \mathcal{E} \tag{2.25}$$

$$\sum_{e \in \mathcal{E}} b_e x_e \leq B \tag{2.26}$$

$$y_e \in \{0,1\} \qquad\qquad e \in \mathcal{E} \tag{2.27}$$

$$x_e \in \{0,1\} \qquad\qquad e \in \mathcal{E} \tag{2.28}$$

where the objective function (2.23) maximizes the shortest path used by the follower, constraints (2.24) enforce the flow balance conditions, constraints (2.25) ensure that the follower does not use an arc that has been interdicted, constraint (2.26) defines the interdiction budget, and constraints (2.27) and (2.28) define the domain of the variables. One of the approaches that is commonly used to solve these problems is to reformulate the model by replacing the inner problem (i.e., the followers problem) by its dual version. The result is a bilinear maximization problem that can be solved via standard linearization techniques.

A path-based mathematical formulation was presented in [22]. In this paper, the authors used as the connectivity measure the weighted sum of the pairwise connections. For this formulation, a parameter r_{ij} is defined as the weight of the connection between i and j. $\mathcal{P}_{ij}$ is defined as the set of all possible paths connecting the pair of nodes (i,j), and $\mathcal{E}(P)$ is the set of arcs that comprise path $P \in \mathcal{P}_{ij}$. Furthermore, x_{ij} is a binary variable that takes the value of one if nodes i and j are not connected and zero otherwise, and variable y_e is a binary variable that takes the value of one if node i is not deleted in the optimal solution and zero otherwise. The mathematical formulation follows:

$$\max \sum_{i \in \mathcal{V}} r_{ij} x_{ij} \tag{2.29}$$

$$\text{s.t.} \sum_{e \in \mathcal{E}} y_e \leq B \qquad\qquad \forall i \in \mathcal{V} \tag{2.30}$$

$$\sum_{e \in \mathcal{E}(P)} y_e \geq x_{ij} \qquad\qquad \forall i, j \in \mathcal{V}, P \in \mathcal{P}_{ij} \qquad (2.31)$$

$$y_e \in \{0,1\} \qquad\qquad e \in \mathcal{E} \qquad (2.32)$$

$$x_{ij} \in \{0,1\} \qquad\qquad i, j \in \mathcal{V} \qquad (2.33)$$

where the objective function (2.29) maximizes the weighted sum of pairwise connections that are disrupted, constraint (2.30) ensures that no more than B arcs are eliminated, constraints (2.31) ensure that if there is a path between nodes i and j, variable $x_{ij} = 0$. Finally constraints (2.32) and (2.33) define the domain of the variables used. An alternative formulation regarding the critical arc detection problem in this context can be found in [21].

2.4.3 Approximation Algorithms

Dinh et al. [9] presented also an approximation algorithm for detecting critical arcs. Similarly to the approach designed for the CNP, the authors propose to minimize the number of arcs that must be removed in order to achieve a disruption level in the number of the residual pairwise connections. The authors provide a proof of the $\mathcal{NP}$-completeness of this problem by a reduction from the balanced cut problem [10]. Finally, they propose a $O(\log^{1.5})$ pseudo-approximation scheme.

2.5 The Critical Clique Detection Problem (CCP)

The increasing interest on the CNP has motivated recent studies regarding the detection of other critical substructures such as cliques and paths. The CCP has been recently explored in the work of Walteros and Pardalos [29], who examine the extension of some of the approaches originally conceived for the CNP, to tackle the CCP. In their work, the authors prove that the CCP is $\mathcal{NP}$-complete for the general case using a transformation from the partition into cliques problem [11]. They also propose a mathematical formulation as well as a decomposition approach.

The CCP can be defined as follows: Given a connected undirected network $\mathcal{G}(\mathcal{V}, \mathcal{E})$ where $\mathcal{V}$ and $\mathcal{E}$ are the set of nodes and arcs, respectively, and an integer k, the CCP involves finding a set of k disjoint cliques such that its deletion results in the maximum network disconnection. Additional constraints regarding the structure of the cliques can also be imposed, for instance, an upper bound on the size of the critical cliques. Notice that the CCP can be seen as a generalization of the CNP, where the objective is to find cliques instead nodes. The CNP is then the case where the size of the cliques is limited to be one. Figure 2.3 presents an example of the

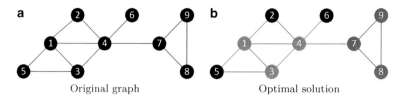

Original graph   Optimal solution

Fig. 2.3 Example for a 9-node graph. (**a**) Original graph (**b**) Optimal solution

CCP over a 9-node graph, where $k = 2$. Figure 2.3a displays the original network, and Fig. 2.3b the optimal solution where the cliques selected are colored in green and blue, respectively.

Among the different connectivity measure described above, the authors discussed two: the total pairwise connections and the size of the largest component. A description of how these measures are incorporated in the mathematical formulations follows:

For any subset $V' \subseteq V$, set $\mathcal{E}(V') \subseteq \mathcal{E}$ is defined as the set of arcs such that, for each arc $e \in \mathcal{E}(V')$, both endpoints of e belong to V'. Let $\mathcal{G}(V')$ be the graph comprised by the set of nodes V', and the set of arcs $\mathcal{E}(V')$. Assume that nodes $i, j \in V$ are connected over $\mathcal{G}$ if there exists at least one path that connects i with j in $\mathcal{G}$. Let $\mathcal{Q}$ be the set of maximal connected components of $\mathcal{G}$. That is, a subset $\mathcal{C}_q \subseteq V$ of nodes such that every pair of nodes $i, j \in \mathcal{C}_q$ is connected over $\mathcal{G}(\mathcal{C}_q)$, and such that, for every node $l \in V \setminus \mathcal{C}_q$, there is no arc connecting l with any node $i \in \mathcal{C}_q$. Let $\sigma_q = |\mathcal{C}_q|$ be the number of nodes of component $\mathcal{C}_q \in \mathcal{Q}$. The number of pairwise connections of component $\mathcal{C}_q \in \mathcal{Q}$ is then defined as $\sigma_q(\sigma_q - 1)/2$. Let $\mathcal{T} = \{\mathcal{K}_1, \ldots, \mathcal{K}_k\}$ be the set of k critical cliques of $\mathcal{G}$, $V(\mathcal{K}_t) \subseteq V$ be the subset of nodes that comprise clique $\mathcal{K}_t \in \mathcal{T}$, and $V(\mathcal{T}) \subseteq V$ be the set of all the nodes that belong to the critical cliques. Finally, let $\mathcal{G}^{\mathcal{T}} = (V \setminus V(\mathcal{T}), \mathcal{E}(V \setminus V(\mathcal{T})))$ be the resulting network after the deletion of the critical cliques, and $\mathcal{Q}^{\mathcal{T}}$ the corresponding set of remaining components. Then, the objectives are:

Minimize the total pairwise connections: Given a network $\mathcal{G} = (V, \mathcal{E})$ and an integer k, find a collection of cliques $\mathcal{T}$, of size $|\mathcal{T}| \leq k$, such that the sum of the pairwise connections of all the components left is minimized:

$$\min \sum_{q \in \mathcal{Q}^{\mathcal{T}}} \sigma_q(\sigma_q - 1)/2 \tag{2.34}$$

Minimize the size of the largest component: Given a network $\mathcal{G} = (V, \mathcal{E})$ and an integer k, find a collection of cliques $\mathcal{T}$, of size $|\mathcal{T}| \leq k$, such that the size of the largest component is minimized:

$$\min \max_{q \in \mathcal{Q}^{\mathcal{T}}} \{\sigma_q\} \tag{2.35}$$

2.5.1 *Mathematical Formulations*

One of formulations described in [29] is as follows. Let $\mathcal{V}(e)$ be the set of endpoints of arc $e \in \mathcal{E}$ and $\mathcal{T}$ be the set of critical cliques such that $|\mathcal{T}| = k$. Let v_i^t be a binary variable that takes the value of one if node i is assigned to clique $\mathcal{K}_t \in \mathcal{T}$. Let u_{ij} be a binary variable that takes the value of one if nodes i and j, belong to the same component and zero otherwise. Let z_i be an auxiliary binary variable that takes the value of one if node i belongs to a component in the residual graph and zero otherwise. The mathematical formulation for the CCP for the TPW objective measure is as follows:

$$\min \sum_{i,j \in \mathcal{V}} u_{ij} \tag{2.36}$$

$$\text{s.t.} \sum_{t \in \mathcal{T}} v_i^t \leq 1 \qquad\qquad i \in \mathcal{V} \tag{2.37}$$

$$v_i^t + v_j^t \leq 1 \qquad\qquad e \in \mathcal{V} \times \mathcal{V} \setminus \mathcal{E}, i, j \in \mathcal{V}(e), t \in \mathcal{T} \tag{2.38}$$

$$z_i + \sum_{t \in \mathcal{T}} v_i^t = 1 \qquad\qquad i \in \mathcal{V} \tag{2.39}$$

$$u_{ij} \geq z_i + z_j - 1 \qquad\qquad e \in \mathcal{E}, i, j \in \mathcal{V}(e) \tag{2.40}$$

$$u_{ij} + u_{jl} - u_{il} \leq 1 \qquad\qquad i, j, l \in \mathcal{V} \tag{2.41}$$

$$u_{ij} - u_{jl} + u_{il} \leq 1 \qquad\qquad i, j, l \in \mathcal{V} \tag{2.42}$$

$$-u_{ij} + u_{jl} + u_{il} \leq 1 \qquad\qquad i, j, l \in \mathcal{V} \tag{2.43}$$

$$v_i^t \in \{0,1\} \qquad\qquad i \in \mathcal{V}, t \in \mathcal{T} \tag{2.44}$$

$$z_i \in \{0,1\} \qquad\qquad i \in \mathcal{V} \tag{2.45}$$

$$u_{ij} \in \{0,1\} \qquad\qquad i, j \in \mathcal{V} \tag{2.46}$$

where the objective function (2.36) minimizes the sum of pairwise connections. Note that since u_{ij} is equal to 1 if nodes i and j belong to the same component, $\sum_{i,j \in \mathcal{V}} u_{ij}$ is equivalent to $\sum_{q \in \mathcal{Q}} \sigma_q(\sigma_q - 1)$. Constraint (2.37) ensures that each node is assigned to at most one clique. Constraint (2.38) ensures that if there is no arc $e \in \mathcal{E}$ between nodes i and j (i.e., $e \in \mathcal{V} \times \mathcal{V} \setminus \mathcal{E}$), both nodes cannot be assigned to the same clique. Constraint (2.39) ensures that if node i is not assigned to a clique, its corresponding variable z_i must be equal to one. Constraints (2.40) define the relationship between u variables and z variables. Constraints (2.41) and (2.42) define the triangular relationship of u variables. And finally constraints (2.44)–(2.46) define the domain of the variables used.

Furthermore, to use the maximum size of the largest component as the objective, above model is adapted by introducing a new variable β defined as the size of the largest component. Then the model is then formulated as follows:

$$\min \quad \beta \tag{2.47}$$

$$\text{s.t.} \quad (2.37 - 2.46)$$

$$\sum_{i \in V} u_{ij} \leq \beta \qquad\qquad i \in V \tag{2.48}$$

where objective function (2.47) combined with constraints (2.48) enforces the minimization of the size of the largest component.

Similarly to the case of the CNP, these two formulations grow relatively large in size with respect to $|\mathcal{V}|$ (they require $O(|\mathcal{V}|^2)$ variables and $O(|\mathcal{V}|^3)$ constraints). To efficiently solve these formulations, it is common to use a cutting plane generation scheme that sequentially includes constraints (2.3)–(2.5) as needed. Moreover, it is easy to see that we can strengthen these formulations using some valid inequalities originally designed for similar problems [14, 23], as well as symmetry-breaking constraints.

2.6   Concluding Remarks and Further Directions

This study was motivated by the increasing interest of solving critical element detection problems on analyzing graphs. Recently, several research efforts have been put together to develop approaches and techniques to efficiently solve these kinds of problems. In this paper we outline and relate these approaches and survey mainly recent contributions.

Most of the extant literature has been focused on identifying critical nodes and critical arcs. Many techniques and applications have been published regarding these two problems and their applications. However, because of the recent nature of these problems, there are still plenty of different trends that can be followed to improve the current solution techniques.

For example, the use of other metaheuristics such as genetic algorithms, tabu search, or ant colony schemes has yet to be explored. These techniques can be very fruitful to obtain solutions for large-scale instances, generally out of reach for exact solution approaches.

Another possible path may involve extending some of the available techniques that were originally tailored to solve particular cases of these problems, to tackle more general versions. This could be the case of the dynamic programming schemes initially designed to solve CNPs over trees.

Additionally, in contrast to the current state of the art regarding the detection of critical nodes and arcs, there is still very few attention to the identification of more complex critical structures (paths, cliques, clusters, etc.), despite the evident applicability. We believe that this is one of the possible research paths to follow over the next years.

Finally, a future task will also involve the application of critical element detection problems in new fields such as neuroscience, biology, and genetics.

References

1. Ahuja, R.K., Magnanti, T.L., Orlin, J.B.: Network Flows: Theory, Algorithms, and Applications. Prentice-Hall, Inc., Upper Saddle River, NJ, USA (1993)
2. Albert, R., Jeong, H., Barabasi, A.-L.: Error and attack tolerance of complex networks. Nature **406**(6794), 378–382 (2000)
3. Arulselvan, A., Commander, C.W., Elefteriadou, L., Pardalos P.M.: Detecting critical nodes in sparse graphs. Comput. Oper. Res. **36**(7), 2193–2200 (2009)
4. Borgatti, S.P.: Identifying sets of key players in a social network. Comput. Math. Organ. Theor. **12**, 21–34 (2006)
5. Borgatti, S.P., Everett, M.G.: A graph-theoretic perspective on centrality. Soc. Network **28**(4), 466–484 (2006)
6. Church, R.L., Scaparra, M.P., Middleton, R.S.: Identifying critical infrastructure: The median and covering facility interdiction problems. Ann. Assoc. Am. Geogr. **94**(3), 491–502 (2004)
7. Corley, H., Sha, D.Y.: Most vital links and nodes in weighted networks. Oper. Res. Lett. **1**(4), 157–160 (1982)
8. Di Summa, M., Grosso, A., Locatelli, M.: Complexity of the critical node problem over trees. Comput. Oper. Res. **38**(12), 1766–1774 (2011)
9. Dinh, T.N., Xuan, Y., Thai, M.T., Pardalos, P.M., Znati, T.: On new approaches of assessing network vulnerability: Hardness and approximation, Networking, IEEE/ACM Transactions on, 20(2), 609–619 (2012)
10. Garey, M., Johnson, D., Stockmeyer, L.: Some simplified np-complete graph problems. Theor. Comput. Sci. **1**(3), 237–267 (1976)
11. Garey, M.R., Johnson, D.S.: Computers and Intractability: A Guide to the Theory of NP-Completeness. W. H. Freeman & Co., New York, NY, USA (1990)
12. Garg, N., Vazirani, V., Yannakakis, M.: Primal-dual approximation algorithms for integral flow and multicut in trees. Algorithmica **18**, 3–20 (1997)
13. Grötschel, M., Monma, C., Stoer, M.: Design of survivable networks. In: Ball, C.M.M.O., Magnanti, T.L., Nemhauser, G. (eds.) Network Models. Handbooks in Operations Research and Management Science, vol. 7, pp. 617–672. Elsevier (1995)
14. Grötschel, M., Wakabayashi, Y.: Facets of the clique partitioning polytope. Math. Program. **47**, 367–387 (1990)
15. Grubesic, T.H., Matisziw T.C., Murray, A.T., Snediker D.: Comparative approaches for assessing network vulnerability. Int. Reg. Sci. Rev. **31**(1), 88–112 (2008)
16. T. H. Grubesic and A. T. Murray. Vital nodes, interconnected infrastructures, and the geographies of network survivability. Ann. Assoc. Am. Geogr. **96**(1), 64–83 (2006)
17. Houck, D.J., Kim, E., O'Reilly, G.P., Picklesimer, D.D., Uzunalioglu, H.: A network survivability model for critical national infrastructures. Bell Labs Technical J. **8**(4), 153–172 (2004)
18. Israeli, E., Wood, R.K.: Shortest-path network interdiction. Networks **40**(2), 97–111 (2002)
19. Jenelius, E., Petersen, T., Mattsson L.-G.: Importance and exposure in road network vulnerability analysis. Transport. Res. Part A: Pol. Prac. **40**(7), 537–560 (2006)
20. Lim, C., Smith, J.C.: Algorithms for discrete and continuous multicommodity flow network interdiction problems. IIE Transactions **39**(1), 15–26 (2007)

21. Matisziw, T.C., Murray, A.T.: Modeling s-t path availability to support disaster vulnerability assessment of network infrastructure. Comput. Oper. Res. **36**, 16–26 (2009)
22. Myung, Y.-S., joon Kim, H.: A cutting plane algorithm for computing k-edge survivability of a network. Eur. J. Oper. Res. **156**(3), 579–589 (2004)
23. Oosten, M., Rutten, J.H.G.C., Spieksma, F.C.R.: Disconnecting graphs by removing vertices: a polyhedral approach. Statistica Neerlandica **61**(1), 35–60 (2007)
24. Salmeron, J., Wood, K.R., Baldick, R.: Analysis of electric grid security under terrorist threat. IEEE Transactions on Power Systems. **19**(2), 905–912 (2004)
25. Shen, S., Smith, J.C.: Polynomial-time algorithms for solving a class of critical node problems on trees and series-parallel graphs. Networks, Wiley Subscription Services, Inc., A Wiley Company (2011) doi: 10.1002/net.20464
26. Shmoys, D.B.: Cut problems and their application to divide-and-conquer, pages 192–235. PWS Publishing Co., Boston, MA, USA (1997)
27. Simaan, M., Cruz, J.B.: On the stackelberg strategy in nonzero-sum games. J. Optim. Theor. Appl. **11**, 533–555 (1973)
28. Tao, Z., Zhongqian, F., Binghong, W.: Epidemic dynamics on complex networks. Progr. Nat. Sci. **16**(5), 452–457 (2006) doi: 10.1080/10020070612330019
29. Walteros, J.L., Pardalos, P.M.: A decomposition approach for solving critical clique detection problems, Experimental Algorithms. In: Klasing, R., (ed.) Lecture Notes in Computer Science, vol. **7276**, pp. 393–404, Springer Berlin, Heidelberg (2012)
30. Wollmer, R.: Removing arcs from a network. Oper. Res. **12**(6), 934–940 (1964)
31. Wood, R.K.: Deterministic network interdiction. Math. Comput. Model. **17**(2), 1–18 (1993)

Chapter 3
Study of Engagement with Mobile Targets

Spiridon Tassopoulos

Abstract This study intends to describe a methodology and to provide the data required for a realistic analysis of a mobile target engagement. Specifically, it provides a means of estimating the probability that a target is still present at an observed location as a function of time from the observation where the time the target stopped is unknown. With this methodology targets can then be evaluated not only on the basis of expected fractional coverage as in the manual, but also on the basis of whether there is an adequate likelihood that they will still be present when weapon arrives.

Keywords Engagement • Acquisition • Mobile target • Dwell time • Renewal Theory

Mathematics Subject Classification (2010): 26A42, 60H30, 60K05, 65C20, 65C30

3.1 Introduction

The scope of this paper is to describe a methodology and to provide the data required for a realistic analysis of a mobile target engagement ([2–4]). More specifically, it provides a means of estimating the probability that a target is still present at an observed location as a function of time from the observation where the time the target stopped is unknown.

S. Tassopoulos (✉)
Eleftherias 1, Nea Kios Argolidas, GR 21053, Argolida, Greece
e-mail: spitasso@yahoo.gr

N.J. Daras (ed.), *Applications of Mathematics and Informatics in Military Science*,
Springer Optimization and Its Applications 71, DOI 10.1007/978-1-4614-4109-0_3,
© Springer Science+Business Media New York 2012

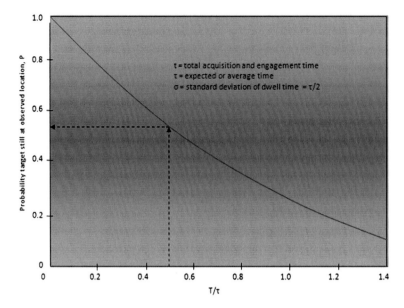

Fig. 3.1 Probability of target presence

Mobile target is defined both as a target that moves nearly continuously (such as a tank company) and as one that moves only occasionally (such as an artillery battery or command post).

With this methodology targets can then be evaluated not only on the basis of expected fractional coverage as in the manuals but also on the basis of whether there is an adequate likelihood that they will still be present when weapon arrives. Results are summarized in Fig. 3.1.

This chart shows the probability of target being present at an observed location as a function of the expected target dwell time (τ) and the acquisition/engagement time (t). Its use is best illustrated by an example. Let us assume an expected target dwell time (τ) of 12 h and that the time (t) necessary to acquire and process the target information, to communicate it to required elements, to make decisions, to plan and prepare weapon use, and to deploy the weapon is 6 h. The ratio of t to is therefore 0, 5. The resultant expected probability is about 0, 53. Thus, there is slightly better than a 50/50 change of the target still being present when the weapon actually arrives for example (Fig. 3.2).

We begin by assuming that the probability of the target leaving its original position between t and $t + DT$ is

$$P_1(t)dt = \frac{1}{C\sigma\sqrt{2\pi}}e^{-\frac{(t-\tau)^2}{2\sigma^2}}\,dt$$

Fig. 3.2 Target dwell time distribution model

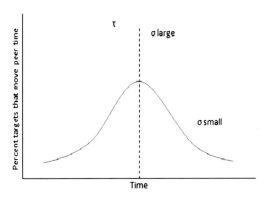

where $t = 0$ is the time at which the target originally settled into the given position, τ is the average time that the target remains in place, σ^2 is the variance in the distribution, and the normalization constant

$$C = \frac{1}{2} + \frac{1}{2}\mathrm{erf}\left(\frac{\tau}{\sigma\sqrt{2}}\right)$$

is chosen such that

$$\int_0^{+\infty} P_1(t)\,dt = 1.$$

66% of the target will leave between $\tau - \sigma$ and $\tau + \sigma$.
95% of the target will leave between $\tau - 2\sigma$ and $\tau + 2\sigma$.
We now assume that the target is detected at some arbitrary time $t = t_1$ and we wish to know the probability destiny ($P_2(t_2)$) of the time t_2 between detection and the departure of the target.

This turns out to be one of the main problems of a branch of probability theory called Renewal Theory. The random variable t_2 is called the residual waiting time or the excess lifetime. Using the results of Renewal Theory it can be shown that the probability that the target will leave at a time t_2 after it is detected is

$$P_2(t_2)dt_2 = \frac{1 - F_1(t_2)}{\mu}dt_2$$

where

$$F_1(t_2) = \int_0^{t_2} P_2(t)\,dt$$

And/or, integrating by parts and using $F_1(\infty) = 1$

$$\mu = \int_0^{\infty} tP_1(t)\,dt$$

$$\mu = \int_0^{\infty} (1 - F_1(t))\,dt$$

so that

$$P_2(t_2)dt_2 = \frac{[1 - F_1(t_2)]dt_2}{\int_0^\infty (1 - F_1(t))\,dt}$$

Using the original expression for P_1 (t) we now have

$$1 - F_1(t_2) = \int_{t_2}^\infty P_1(t)\,dt = \frac{1}{C\sigma\sqrt{2\pi}} e^{-\frac{(t-\tau)^2}{2\sigma^2}}\,dt$$

Setting $t - \tau = \sigma\sqrt{2}x$, this becomes

$$1 - F_1(t_2) = \frac{1}{C\sqrt{\pi}} \int_{\frac{t_2-\tau}{\sigma\sqrt{\pi}}}^\infty e^{-x^2}\,dx$$

which gives

$$1 - F_1(t_2) = \frac{1}{2C}\left\{1 - \mathrm{erf}\left[\frac{t_2 - \tau}{\sigma\sqrt{2}}\right]\right\}$$

where $\mathrm{erf}(x)$ is the error function, so that the expression for $P_2(t_2)dt_2$ becomes

$$P_2(t_2)dt_2 = \frac{\left\{1 - \mathrm{erf}\left[\frac{t_2-\tau}{\sigma\sqrt{2}}\right]\right\}dt_2}{\int_0^\infty \left\{1 - \mathrm{erf}\left[\frac{t-\tau}{\sigma\sqrt{2}}\right]\right\}dt}.$$

This can be further simplified. Using the formulas

$$\int \mathrm{erf}(\alpha x)dx = x\,\mathrm{erf}(\alpha x) + \frac{e^{-\alpha^2 x^2}}{\alpha\sqrt{\pi}}$$

and

$$\int_0^\infty [1 - \mathrm{erf}(\alpha x)]dx = \frac{1}{\alpha\sqrt{\pi}}$$

we obtain

$$\int_y^\infty [1 - \mathrm{erf}(\alpha x)]dx = \frac{e^{-\alpha^2 y^2}}{\alpha\sqrt{\pi}} - y[1 - \mathrm{erf}(\alpha y)].$$

Using this we have

$$\int_0^\infty \left\{1 - \mathrm{erf}\left[\frac{t-\tau}{\sigma\sqrt{2}}\right]\right\}dt = \sigma\sqrt{\frac{2}{\pi}}e^{-\frac{\tau^2}{2\sigma^2}} + \tau\left[1 + \mathrm{erf}\left(\frac{\tau}{\sigma\sqrt{2}}\right)\right].$$

So that the final result for $P_2(t_2)dt_2$ is

$$P_2(t_2)dt_2 = \frac{\left\{1 - \mathrm{erf}\left[\frac{t_2-\tau}{\sigma\sqrt{2}}\right]\right\}dt_2}{\sigma\sqrt{\frac{2}{\pi}}e^{-\frac{\tau^2}{2\sigma^2}} + \tau\left[1 + \mathrm{erf}\left(\frac{\tau}{\sigma\sqrt{2}}\right)\right]}.$$

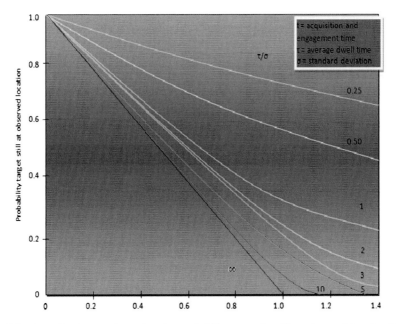

Fig. 3.3 Generalized curve for estimating probability target is still at observed location [1, 5] and [6].

The probability that the target will still be present at a time t_3 after it was detected is now given by

$$P_3(t_3) = 1 - \int_0^{t_3} P_2(t_2)\,dt_2$$

or, since,

$$\int_0^{+\infty} P_2(t_2)\,dt_2 = 1$$

$$P_3(t_3) = \int_{t_3}^{+\infty} P_2(t_2)\,dt_2.$$

Using the expression for P_2, we obtain after integrating

$$P_3(t_3) = \frac{\sigma\sqrt{\frac{2}{\pi}}e^{-\frac{\pi^2}{e2\sigma^2}} - (t_3 - \tau)\left\{1 - \mathrm{erf}\left[\frac{t_3-\tau}{\sigma\sqrt{2}}\right]\right\}}{\sigma\sqrt{\frac{2}{\pi}}e^{-\frac{\pi^2}{e2\sigma^2}} + \tau\left[1 + \mathrm{erf}\left(\frac{\tau}{\sigma\sqrt{2}}\right)\right]}$$

Figure 3.3 presents the generalized curve for the probability of a target being present as a function of time from detection. The probability is presented as a function of two parameters, $\frac{t}{\sigma}$ and $\frac{t}{\tau}$, where τ is the average target dwell time, σ denotes standard deviation of the dwell time, and t denotes the time.

References

1. Abramowitz, M., Stegun, I.A.: Handbook of Mathematical Functions. Dover Publications Inc., New York (1965)
2. U.S. Department of Defence/Department of the Army/etc. Department of the Army/FM 101-31-1
3. U.S. Department of Defence/Department of the Army/etc. Department of the Army/Nuclear Weapons Employment Doctrine and Procedures
4. U.S. Department of Defence/Department of the Army/etc. Department of the Army/Staff Officers Field Manual
5. Feller, W.: An Introduction to Probability Theory and its Applications. vol. 2, 2nd edn, John Willey and Sons, New York (1971)
6. Gradshteyn, I.W., Ryzhik, I.M.: Table of Integrals Series and Products. Academic Press, New York (1965)

Part III
Signal Processing, Scattering

Chapter 4
Solving an Electromagnetic Scattering Problem in Chiral Media

Christodoulos Athanasiadis, Sotiria Dimitroula,
and Kostantinos Skourogiannis

Abstract In this work we consider the problem of scattering of a plane electromagnetic wave by a chiral dielectric obstacle in a chiral environment. We formulate the problem in terms of Beltrami fields in order to state existence and uniqueness. We prove a general scattering theorem when the incident field is a chiral electromagnetic Herglotz pair. Using low-frequency techniques the scattering problem is reduced to an iterative sequence of potential problems which can be solved successively in terms of expansions in appropriate ellipsoidal harmonic functions and we evaluate the zeroth-order approximation.

Keywords Chiral media • Herglotz functions • Far-field operator • Low-frequency scattering

Mathematics Subject Classification (2010): 35P25, 35Q60, 78A40

4.1 Introduction

Chiral materials exhibit the phenomenon of optical activity, that is, the plane of vibration of linearly polarized light is rotated upon passing through an optically active medium. Arago (1811) and Biot (1812), first, studied optically active materials. Pasteur (1848, 1850) found that the arrangement of atoms within a molecule of a natural optically active material is asymmetric in having a nonsuperposable mirror image, but Kelvin was the one to introduce the term chirality, which comes from the greek word "hand" ([25], pp.83–89). Chirality is common in a variety of naturally occurring and man-made objects, such as the DNA molecular scale and helices.

C. Athanasiadis (✉) • S. Dimitroula • K. Skourogiannis
Department of Mathematics, University of Athens, Panepistemiopolis 15784, Athens, Greece
e-mail: cathan@math.uoa.gr; sodimitr@math.uoa.gr; skouroco@otenet.gr

N.J. Daras (ed.), *Applications of Mathematics and Informatics in Military Science*,
Springer Optimization and Its Applications 71, DOI 10.1007/978-1-4614-4109-0_4,
© Springer Science+Business Media New York 2012

It has applications in medicine as one-third of all medical drugs are characterized by chirality. It has also applications in military aircraft as well as in detection of targets by radars [4]. Chirality is introduced into the classical Maxwell equations by a pair of generalized constitutive relations in which the electric and magnetic field are coupled via a new material parameter. In a homogeneous isotropic chiral medium the electromagnetic fields are composed of left-circularly polarized (LCP) and right-circularly polarized (RCP) components. These components have different wave numbers and independent directions of propagation. When either an LCP or a RCP (or a linear combination of LCP and RCP) electromagnetic wave is incident upon a chiral scatterer then the scattered field is composed of both LCP and RCP components. This leads to the derivation of both LCP and RCP far-field patterns.

In recent years chiral materials have been studied intensively in the electro-magnetic theory literature; indicatively we refer to the books written by Lakhtakia (1990), Lakhtakia [25], Lakhtakia et al. [24] and Lindell et al. [27]. Representative of works following rigorous mathematical analysis in the study of chiral media have given by Ammari et al. in [1, 2] and [3]. In [8, 9] and [28] the existence and uniqueness of electromagnetic wave-scattering problems by chiral obstacles in chiral media, using the boundary integral method, have been given. Scattering relations for electromagnetic waves in chiral media have been studied in [11] for planes waves and in [14] for spherical waves. Moreover, Beltrami Herglotz functions for electromagnetic chiral media have been defined in [12] and [13].

In this paper we consider the scattering of time-harmonic waves by a bounded three-dimensional homogeneous penetrable chiral obstacle, with homogeneous host, embedded in a chiral environment. In the second part, we formulate the scattering problem and present the Bohren decomposition [15, 16] of the electric and magnetic fields into left-handed and right-handed Beltrami fields [8,9] and [28]. More specifically, we restate the scattering problem in terms of Beltrami fields and we formulate an equivalent to the scattering problem integral equation. In the third part, we define the LCP and RCP electric far-field patterns via the asymptotic behavior of the scattered electric field [11]. We state general scattering theorems by which we obtain closed form expressions for the scattering cross sections in terms of the forward LCP or RCP far-field pattern. We state some results on the LCP, RCP far-field operators [12] and [13]; considering as an incident field, a superposition of incident fields, we prove the General Scattering Theorem for both LCP and RCP operators. These results can be used in solving inverse scattering problems in chiral media. Finally, we give the integral representation of the solution of the transmission problem as well as some results on the approximation of the zeroth order of the solution via the method of low-frequency scattering in ellipsodial geometry [2,6,9] and [29].

A homogeneous isotropic chiral medium in a region D is characterized by three parameters, the electric permittivity ε, the magnetic permeability μ, and the chirality measure β. If $\mathbf{E}, \mathbf{H}$ are the electric and the magnetic fields and considering a time dependence of $e^{-i\omega t}$, $\omega > 0$ being the angular frequency, throughout, we have in a source free region

$$\nabla \times \mathbf{E} - i\omega\mathbf{B} = \mathbf{0}, \quad \nabla \times \mathbf{H} + i\omega\mathbf{D} = \mathbf{0}. \tag{4.1}$$

The chirality measure β appears in the Drude-Born-Fedorov constitutive relations that we make use of in this paper

$$\mathbf{D} = \varepsilon(\mathbf{E} + \beta\nabla \times \mathbf{E}), \quad \mathbf{B} = \mu(\mathbf{H} + \beta\nabla \times \mathbf{H}). \tag{4.2}$$

Moreover, we have that $\mathbf{E}$ and $\mathbf{H}$ are divergence-free, that is $\nabla \cdot \mathbf{E} = 0$, $\nabla \cdot \mathbf{H} = 0$. Hence, eliminating $\mathbf{B}$ and $\mathbf{D}$, we get

$$\nabla \times \mathbf{E} = \gamma^2 \beta \mathbf{E} + i\omega\mu \left(\frac{\gamma}{\kappa}\right)^2 \mathbf{H}, \tag{4.3}$$

$$\nabla \times \mathbf{H} = \gamma^2 \beta \mathbf{H} - i\omega\varepsilon \left(\frac{\gamma}{\kappa}\right)^2 \mathbf{E}, \tag{4.4}$$

where $\kappa = \omega\sqrt{\varepsilon\mu}$ is simply a shorthand notation, not a wave number, and $\gamma^2 = \kappa^2(1 - \kappa^2\beta^2)^{-1}$. We always assume that $|\kappa\beta| < 1$.

In chiral media, left-handed and right-handed waves can both propagate independently and with different phase speeds. To see this, we consider the Bohren decomposition of $\mathbf{E}$, $\mathbf{H}$ into suitable Beltrami fields $\mathbf{Q}_L$, $\mathbf{Q}_R$

$$\mathbf{E} = \mathbf{Q}_L + \mathbf{Q}_R, \quad \mathbf{H} = \frac{1}{i\eta}(\mathbf{Q}_L - \mathbf{Q}_R), \tag{4.5}$$

where $\eta = \sqrt{\frac{\mu}{\varepsilon}}$ is the intrinsic impedance of the chiral medium, and

$$\nabla \times \mathbf{Q}_L = \gamma_L \mathbf{Q}_L, \quad \nabla \times \mathbf{Q}_R = -\gamma_R \mathbf{Q}_R, \tag{4.6}$$

with

$$\gamma_L = \kappa(1 - \kappa\beta)^{-1}, \quad \gamma_R = \kappa(1 + \kappa\beta)^{-1}, \tag{4.7}$$

being the wave numbers of $\mathbf{Q}_L$, $\mathbf{Q}_R$, respectively.

4.2   The Scattering Problem

We will proceed by formulating the scattering problem. Let D_1 denote a bounded three-dimensional domain with a smooth closed boundary, S, and connected exterior, D, where D_1 and D are filled with different chiral media of parameters $\varepsilon_1, \mu_1, \beta_1$ and ε, μ, β, respectively. Consider an incident electromagnetic field upon the obstacle D_1,

$$\mathbf{E}^{\text{inc}}(\mathbf{r}) = \mathbf{b}_L e^{i\gamma_L \hat{\mathbf{d}}_L \cdot \mathbf{r}} + \mathbf{b}_R e^{i\gamma_R \hat{\mathbf{d}}_R \cdot \mathbf{r}},$$

$$\mathbf{H}^{\text{inc}}(\mathbf{r}) = \frac{1}{i\eta}\left(\mathbf{b}_L e^{i\gamma_L \hat{\mathbf{d}}_L \cdot \mathbf{r}} - \mathbf{b}_R e^{i\gamma_R \hat{\mathbf{d}}_R \cdot \mathbf{r}}\right) \tag{4.8}$$

where the following relations hold for the polarization vectors $\mathbf{b}_L$, $\mathbf{b}_R$ and the propagation unit vectors $\widehat{\mathbf{d}}_L$, $\widehat{\mathbf{d}}_R$

$$\mathbf{b}_L \cdot \widehat{\mathbf{d}}_L = 0, \; \mathbf{b}_L \times \widehat{\mathbf{d}}_L = -i\mathbf{b}_L,$$

$$\mathbf{b}_R \cdot \widehat{\mathbf{d}}_R = 0, \; \mathbf{b}_R \times \widehat{\mathbf{d}}_R = i\mathbf{b}_R. \tag{4.9}$$

The electromagnetic incident field is partially scattered and partially transmitted into the obstacle. Note that

$$\nabla \times \mathbf{b}_L e^{i\gamma_L \widehat{\mathbf{d}}_L \cdot \mathbf{r}} = \gamma_L \mathbf{b}_L e^{i\gamma_L \widehat{\mathbf{d}}_L \cdot \mathbf{r}},$$

$$\nabla \times \mathbf{b}_R e^{i\gamma_R \widehat{\mathbf{d}}_R \cdot \mathbf{r}} = -\gamma_R \mathbf{b}_R e^{i\gamma_R \widehat{\mathbf{d}}_R \cdot \mathbf{r}},$$

from which we conclude that the incident field $\mathbf{E}^{\mathrm{inc}}(\mathbf{r})$ is a combination of the LCP incident plane wave and the RCP incident plane wave. This leads to the following transmission problem.

Find electric fields $\mathbf{E}_1$, $\mathbf{E}$, and magnetic fields $\mathbf{H}_1$, $\mathbf{H}$ that satisfy the following modified Maxwell's equations:

$$\nabla \times \mathbf{E}_1 = \gamma_1^2 \beta_1 \mathbf{E}_1 + i\omega_1 \mu_1 \left(\frac{\gamma_1}{\kappa_1}\right)^2 \mathbf{H}_1 \text{ in } D_1,$$

$$\nabla \times \mathbf{H}_1 = \gamma_1^2 \beta_1 \mathbf{H}_1 - i\omega_1 \varepsilon_1 \left(\frac{\gamma_1}{\kappa_1}\right)^2 \mathbf{E}_1 \text{ in } D_1, \tag{4.10}$$

$$\nabla \times \mathbf{E} = \gamma^2 \beta \mathbf{E} + i\omega \mu \left(\frac{\gamma}{\kappa}\right)^2 \mathbf{H} \text{ in } D,$$

$$\nabla \times \mathbf{H} = \gamma^2 \beta \mathbf{H} - i\omega \varepsilon \left(\frac{\gamma}{\kappa}\right)^2 \mathbf{E} \text{ in } D. \tag{4.11}$$

The transmission conditions on the interface, S, of, D_1 are

$$\widehat{\mathbf{n}} \times \mathbf{E}^t = \widehat{\mathbf{n}} \times \mathbf{E}_1, \;\; \widehat{\mathbf{n}} \times \mathbf{H}^t = \widehat{\mathbf{n}} \times \mathbf{H}_1 \text{ on } S, \tag{4.12}$$

where $\widehat{\mathbf{n}}$ is the unit outward normal to S and the total exterior fields are given by

$$\mathbf{E}^t = \mathbf{E} + \mathbf{E}^{\mathrm{inc}}, \;\; \mathbf{H}^t = \mathbf{H} + \mathbf{H}^{\mathrm{inc}} \text{ in } D. \tag{4.13}$$

The scattered fields satisfy the following Silver-Müller radiation conditions:

$$\widehat{\mathbf{r}} \times \mathbf{H}(\mathbf{r}) + \eta^{-1} \mathbf{E}(\mathbf{r}) = o\left(\frac{1}{r}\right), \quad r \to \infty, \tag{4.14}$$

$$\widehat{\mathbf{r}} \times \mathbf{E}(\mathbf{r}) - \eta \mathbf{H}(\mathbf{r}) = o\left(\frac{1}{r}\right), \quad r \to \infty, \tag{4.15}$$

uniformly for all directions $\widehat{\mathbf{r}} = \frac{\mathbf{r}}{r}$, where $r = |\mathbf{r}|$. The physical parameters ε_1, μ_1, β_1, γ_{1L}, γ_{1R}, and γ_1 are defined and connected with each other as in the region D adding the index "1." For this problem, in what follows, we will prove a "general scattering theorem" for LCP, RCP Herglotz incident fields [12, 20, 21] that is important for solving inverse scattering problems. Moreover, we will apply low-frequency theory for chiral media and we reduce the problem in a sequence of potential theory problems. We also calculate the zeroth-order approximation in ellipsoidal geometry.

This problem is well posed and its solvability has been studied extensively in [7, 8] using the boundary integral method. Here, we will give a brief description of this method that will help us in the third part of this work.

First we are going to define the function spaces we will employ. Let X be the smooth boundary of an open set in $\mathbb{R}^3$ and let $H^s(X)$ be the L^2-based Sobolev spaces on X. If F is a function space on the (smooth) boundary of an open bounded set in $\mathbb{R}^3$, then TF is the space of all tangential fields with cartesian components in F. We denote $\overline{H}_{loc}^1(D)$ the space of all $\mathbf{u} \in \mathcal{D}'(\mathbb{R}^3)$ such that for all open balls B containing $\overline{D}$, we have $\mathbf{u}|_{B \cap D} \in H^1(B \cap D)$. Let Div be the surface divergence on S. Then,

$$H_{\mathrm{Div}}^1(D_1) = \left\{ \mathbf{u} \in H^1(D) : \quad \mathrm{Div}(\widehat{\mathbf{n}} \times \mathbf{v}) \in H^{1/2}(S) \right\},$$

$$\overline{H}_{\mathrm{loc,Div}}^1(D) = \left\{ \mathbf{u} \in \overline{H}_{\mathrm{loc}}^1(D) : \quad \mathrm{Div}(\mathbf{n} \times \mathbf{v}) \in H^{1/2}(S) \right\},$$

$$TH_{\mathrm{Div}}^{1/2}(S) = \left\{ \mathbf{u} \in TH^{1/2}(D) : \quad \mathrm{Div}(\mathbf{u}) \in H^{1/2}(S) \right\}.$$

We rewrite the transmission problem in terms of Beltrami fields (for details we refer to [7]), and we obtain that the interior LCP (resp. RCP) and the exterior LCP (resp. RCP) satisfy the Helmhlotz equation,

$$\Delta \mathbf{Q}_{1L} + \gamma_{1L}^2 \mathbf{Q}_{1L} = \mathbf{0} \text{ in } D_1, \quad \Delta \mathbf{Q}_{1R} + \gamma_{1R}^2 \mathbf{Q}_{1R} = \mathbf{0} \text{ in } D_1,$$

$$\Delta \mathbf{Q}_L + \gamma_L^2 \mathbf{Q}_L = \mathbf{0} \text{ in } D, \quad \Delta \mathbf{Q}_R + \gamma_R^2 \mathbf{Q}_R = \mathbf{0} \text{ in } D.$$

Hence we may use the classical layer potentials. For $\kappa \in \mathbb{C}$ with $\mathrm{Im}\, \kappa \geq 0$, we define the single-layer potential,

$$(S_1(\kappa)\upsilon)(\mathbf{x}) = \int_S \upsilon(\mathbf{y})\Phi(\mathbf{x}, \mathbf{y}; \kappa)\mathrm{d}s(\mathbf{y}), \mathbf{x} \in D_1, \tag{4.16}$$

where $\mathbf{y} \in S$, $\upsilon(\mathbf{y})$ is a continuous density function and

$$\Phi(\mathbf{x}, \mathbf{y}; \kappa) = \frac{e^{i\kappa|\mathbf{x}-\mathbf{y}|}}{4\pi|\mathbf{x}-\mathbf{y}|}, \mathbf{x} \neq \mathbf{y}$$

is the fundamental solution of the Helmholtz equation.

Consider $\mathbf{a}(\mathbf{y})$ be a tangential vector density, that is $\mathbf{a}(\mathbf{y}) \cdot \widehat{\mathbf{n}}(\mathbf{y}) = 0$ for all $\mathbf{y} \in S$. We define

$$(C_1(\kappa)\mathbf{a})(\mathbf{x}) = \nabla \times \{S_1(\kappa)\mathbf{a}\},$$

and

$$(F_1(\kappa)\mathbf{a})(\mathbf{x}) = \nabla \times \{C_1(\kappa)\mathbf{a}\}.$$

It is known [8, 18] that there exist continuous extensions for the interior and the exterior of the domain D_1

$$C_1(\kappa), F_1(\kappa) : TH_{\mathrm{Div}}^{1/2}(S) \to H_{\mathrm{Div}}^1(D_1),$$

$$C(\kappa), F(\kappa) : TH_{\mathrm{Div}}^{1/2}(S) \to \overline{H}_{\mathrm{loc,Div}}^1(D).$$

It is also known that there exist continuous extensions on the boundary S

$$C(\kappa), F(\kappa) : TH_{\mathrm{Div}}^{1/2}(S) \to TH_{\mathrm{Div}}^{1/2}(S).$$

In addition, $C(\kappa)$ is compact and the following mappings are continuous:

$$C(\kappa) : TH^s(S) \to TH^{s+1}(S), s \in \mathbb{R},$$

and

$$F(\kappa) : TH^s(S) \to TH^{s-1}(S), s \in \mathbb{R}.$$

Moreover, if $v \in TH_{\mathrm{Div}}^{1/2}(S)$, we have for the traces on S,

$$\widetilde{C}(\kappa)v = \widehat{\mathbf{n}} \times C_1(\kappa)v + \frac{1}{2}v = \widehat{\mathbf{n}} \times C(\kappa)v - \frac{1}{2}v, \tag{4.17}$$

$$\widetilde{F}(\kappa)v = \widehat{\mathbf{n}} \times F_1(\kappa)v = \widehat{\mathbf{n}} \times F(\kappa)v. \tag{4.18}$$

We rewrite the solution of the problem as linear combinations of $C_1(\kappa)$, $F_1(\kappa)$, and $C(\kappa)$, $F(\kappa)$, [8–10] and hence we conclude to the following integral equation:

$$(M + K)\widetilde{\phi} = \widetilde{\mathbf{f}}, \tag{4.19}$$

where

$$\widetilde{\phi} = \begin{pmatrix} \phi_1 \\ \phi_2 \end{pmatrix}, \phi_1, \phi_2 \in TH^{1/2}(S), \quad \widetilde{\mathbf{f}} = \begin{pmatrix} \mathbf{f}_1 \\ \mathbf{f}_2 \end{pmatrix}, \mathbf{f}_1, \mathbf{f}_2 \in TH_{\mathrm{Div}}^{1/2}(S),$$

and M, K, are linear combinations of C_1, C, F_1 and F and thus are compact operators. Next, we state a uniqueness result for the transmission problem (4.10)–(4.15), for this, consider the corresponding homogeneous transmission problem, that is the problem arising from (4.10)–(4.15) when $\mathbf{E}^{\mathrm{inc}} = \mathbf{H}^{\mathrm{inc}} = \mathbf{0}$. In [8] the following theorems have been proved.

Theorem 4.1. *The corresponding homogeneous transmission problem has only the trivial solution.*

Theorem 4.2. *If $\widetilde{\phi} \in TH^{1/2}(S) \times TH^{1/2}(S)$ is a solution of the integral equation (4.19), and if $\widetilde{f} \in TH^{1/2}_{Div}(S) \times TH^{1/2}_{Div}(S)$, then $\widetilde{\phi} \in TH^{1/2}_{Div}(S) \times TH^{1/2}_{Div}(S)$.*

4.3   Scattering Relations

In this section we will define the LCP and RCP electric far-field patterns via the asymptotic behavior of the scattered electric field. We will also state general scattering theorems which are very useful in solving the inverse scattering problem and in determining low-frequency expansions for the far-field patterns. We begin our analysis with the following theorem which gives the asymptotic behavior of the scattered electric field.

First, we will write the transmission problem eliminating the magnetic field $\mathbf{H}$, $\mathbf{H}_1$

$$\nabla \times \nabla \times \mathbf{E} - 2\beta\gamma^2 \nabla \times \mathbf{E} - \gamma^2 = \mathbf{0} \quad \text{in } D$$

$$\nabla \times \nabla \times \mathbf{E}_1 - 2\beta_1\gamma_1^2 \nabla \times \mathbf{E}_1 - \gamma_1^2 = \mathbf{0} \quad \text{in } D_1,$$

$$\widehat{\mathbf{n}} \times (\mathbf{E}^{\text{inc}} + \mathbf{E}) = \widehat{\mathbf{n}} \times \mathbf{E}_1$$

$$\frac{\kappa}{\gamma^2\eta} \widehat{\mathbf{n}} \times (\nabla \times (\mathbf{E}^{\text{inc}} + \mathbf{E})) - \frac{\kappa\beta}{\eta} \widehat{\mathbf{n}} \times (\mathbf{E}^{\text{inc}} + \mathbf{E})$$

$$= \frac{\kappa_1}{\gamma_1^2\eta_1} \widehat{\mathbf{n}} \times (\nabla \times \mathbf{E}_1) - \frac{\kappa_1\beta_1}{\eta_1} \widehat{\mathbf{n}} \times \mathbf{E}_1 \qquad \text{on } S,$$

$$\widehat{\mathbf{r}} \times (\nabla \times \mathbf{E}) - 2\beta\gamma^2\widehat{\mathbf{r}} \times \mathbf{E} + \frac{i\gamma^2}{\kappa}\mathbf{E} = o\left(\frac{1}{r}\right), \; r \to \infty, \qquad (4.20)$$

Theorem 4.3. *Let $\mathbf{E}^s \in \mathbf{C}^2(D) \cap \mathbf{C}^1(\overline{D})$ be a solution of the modified Helmholtz equation satisfying the radiation condition. Then $\mathbf{E}^s$ has the asymptotic form*

$$\mathbf{E}^s(\mathbf{r}) = h(\gamma_L r)\mathbf{g}_L(\mathbf{r}) + h(\gamma_R r)\mathbf{g}_R(\mathbf{r}) + O\left(\frac{1}{r^2}\right), \; r \to \infty, \qquad (4.21)$$

uniformly in all directions $\widehat{\mathbf{r}} \in S^2$, where $h(x) = \frac{e^{ix}}{ix}$ is the zeroth-order spherical Hankel function of the first kind. The vector fields $\mathbf{g}_L$ and $\mathbf{g}_R$ are the electric LCP far-field pattern and RCP far-field pattern respectively. They are given by

$$\mathbf{g}_L(\widehat{\mathbf{r}}) = \frac{i\kappa\gamma_L}{8\pi\gamma^2} \widetilde{K}_L(\widehat{\mathbf{r}}) \cdot \int_S \widehat{\mathbf{n}} \times [\gamma_L \nabla \mathbf{E}^s(\mathbf{r}') + \gamma^2 \mathbf{E}^s(\mathbf{r}')] e^{-i\gamma_L \widehat{\mathbf{r}} \cdot \mathbf{r}'} ds(\mathbf{r}') \qquad (4.22)$$

$$\mathbf{g}_R(\hat{\mathbf{r}}) = \frac{i\kappa\gamma_R}{8\pi\gamma^2} \widetilde{K}_R(\hat{\mathbf{r}}) \cdot \int_S \hat{\mathbf{n}} \times [\gamma_R \nabla E^s(\mathbf{r}') - \gamma^2 E^s(\mathbf{r}')] e^{-i\gamma_R \hat{\mathbf{r}} \cdot \mathbf{r}'} \, ds(\mathbf{r}') \qquad (4.23)$$

and satisfy

$$\hat{\mathbf{r}} \cdot \mathbf{g}_L(\hat{\mathbf{r}}) = \hat{\mathbf{r}} \cdot \mathbf{g}_R(\hat{\mathbf{r}}) = 0, \qquad (4.24)$$

$$\hat{\mathbf{r}} \times \mathbf{g}_L(\hat{\mathbf{r}}) = -i\mathbf{g}_L(\hat{\mathbf{r}}), \ \hat{\mathbf{r}} \times \mathbf{g}_R(\hat{\mathbf{r}}) = i\mathbf{g}_R(\hat{\mathbf{r}}). \qquad (4.25)$$

The dyadics $\widetilde{K}_L(\hat{\mathbf{r}})$ and $\widetilde{K}_R(\hat{\mathbf{r}})$ are given by

$$\widetilde{K}_L(\hat{\mathbf{r}}) = \widetilde{I} - \hat{\mathbf{r}}\hat{\mathbf{r}} + i\hat{\mathbf{r}} \times \widetilde{I}, \ \widetilde{K}_R(\hat{\mathbf{r}}) = \widetilde{I} - \hat{\mathbf{r}}\hat{\mathbf{r}} - i\hat{\mathbf{r}} \times \widetilde{I}, \qquad (4.26)$$

where $\widetilde{I} = \hat{\mathbf{x}}\hat{\mathbf{x}} + \hat{\mathbf{y}}\hat{\mathbf{y}} + \hat{\mathbf{z}}\hat{\mathbf{z}}$ is the identity dyadic.

Proof. The scattered electric field has the following integral representation:

$$\mathbf{E}^s(\mathbf{r}) = -2\beta\gamma^2 \int_S \widetilde{B}(\mathbf{r}, \mathbf{r}') \cdot [\hat{\mathbf{n}} \times \mathbf{E}^s(\mathbf{r}')] ds(\mathbf{r}')$$

$$+ \int_S \{\widetilde{B}(\mathbf{r}, \mathbf{r}') \cdot [\hat{\mathbf{n}} \times (\nabla \times \mathbf{E}^s(\mathbf{r}'))]$$

$$+ [\nabla_r \times \widetilde{B}(\mathbf{r}, \mathbf{r}')] \cdot [\hat{\mathbf{n}} \times \mathbf{E}^s(\mathbf{r}')]\} ds(\mathbf{r}'), \qquad (4.27)$$

where $\widetilde{B}(\mathbf{r}, \mathbf{r}')$ is the infinite medium Greeen's dyadic, given by

$$\widetilde{B}(\mathbf{r}, \mathbf{r}') = \widetilde{B}_L(\mathbf{r}, \mathbf{r}') + \widetilde{B}_R(\mathbf{r}, \mathbf{r}'), \qquad (4.28)$$

$$\widetilde{B}_L(\mathbf{r}, \mathbf{r}') = \frac{i\kappa\gamma_L}{8\pi\gamma^2} \left[\gamma_L \widetilde{I} + \frac{1}{\gamma_L} \nabla\nabla + \nabla \times \widetilde{I} \right] h(\gamma_L |\mathbf{r} - \mathbf{r}'|), \qquad (4.29)$$

$$\widetilde{B}_R(r, r') = \frac{i\kappa\gamma_R}{8\pi\gamma^2} \left[\gamma_R \widetilde{I} + \frac{1}{\gamma_R} \nabla\nabla - \nabla \times \widetilde{I} \right] h(\gamma_R |\mathbf{r} - \mathbf{r}'|). \qquad (4.30)$$

Using asymptotic relations, we obtain

$$\widetilde{B}(\mathbf{r}, \mathbf{r}') = \frac{i\kappa\gamma_L}{8\pi\gamma^2} h(\gamma_L r) e^{-i\gamma_L \hat{\mathbf{r}} \cdot \mathbf{r}'} \widetilde{K}_L(\hat{\mathbf{r}}) + \frac{i\kappa\gamma_R}{8\pi\gamma^2} h(\gamma_R r) e^{-i\gamma_R \hat{\mathbf{r}} \cdot \mathbf{r}'} \widetilde{K}_R(\hat{\mathbf{r}})$$

$$+ O(\frac{1}{r^2}), \ r \to \infty \qquad (4.31)$$

$$\nabla_r \times \widetilde{B}(\mathbf{r}, \mathbf{r}') = \frac{i\kappa\gamma_L^3}{8\pi\gamma^2} h(\gamma_L r) e^{-i\gamma_L \hat{\mathbf{r}} \cdot \mathbf{r}'} \widetilde{K}_L(\hat{\mathbf{r}}) - \frac{i\kappa\gamma_R^3}{8\pi\gamma^2} h(\gamma_R r) e^{-i\gamma_R \hat{\mathbf{r}} \cdot \mathbf{r}'} \widetilde{K}_R(\hat{\mathbf{r}})$$

$$+ O(\frac{1}{r^2}), \ r \to \infty \qquad (4.32)$$

Introducing the above asymptotic forms into the integral representation of the electric scattered field we get the integral representations of the LCP and RCP far-field pattern as given in the theorem.   □

Next we are going to state the reciprocity relation for chiral media and the "general scattering theorem." The latter is a connection of the electric far-field patterns for two LCP and two RCP directions and an integral over directions of both LCP and RCP electric far-field patterns.

Theorem 4.4. *Let E_1^i, E_2^i be two plane electric waves incident upon the scatter D_1 and E_1^s, E_1^s the corresponding scattered fields. Then the following reciprocity relation holds true:*

$$\frac{1}{\gamma_L^2} p_{L2} \cdot g_{L1}(-\hat{d}_{L2}) + \frac{1}{\gamma_R^2} p_{R2} \cdot g_{R1}(-\hat{d}_{R2}) = \frac{1}{\gamma_L^2} p_{L1} \cdot g_{L2}(-\hat{d}_{L1}) + \frac{1}{\gamma_L^2} p_{R1} \cdot g_{R2}(-\hat{d}_{R1}).$$

(4.33)

We refer to [11] for the proof.

Theorem 4.5. *Let E_1^i, E_2^i be two plane electric waves incident upon the scatter D_1, E_1^s, E_1^s be the corresponding scattered fields and under the assumption that the transmission conditions are satisfied on S, then the following relation is valid:*

$$\frac{1}{\gamma_L^2} [\overline{p_{L1}} \cdot g_{L2}(\hat{d}_{L1}) + p_{L2} \cdot \overline{g_{L1}(\hat{d}_{L2})}] + \frac{1}{\gamma_R^2} [\overline{p_{R1}} \cdot g_{R2}(\hat{d}_{R1}) + p_{R2} \cdot \overline{g_{R1}(\hat{d}_{R2})}]$$

$$= -\frac{1}{2\pi} \int_{S^2} \left[\frac{1}{\gamma_L^2} \overline{g_{L1}(\hat{r})} \cdot g_{L2}(\hat{r}) + \frac{1}{\gamma_R^2} \overline{g_{R1}(\hat{r})} \cdot g_{R2}(\hat{r}) \right] ds(\hat{r}).$$

(4.34)

We refer to [11] for the proof.
We now consider either an LCP or an RCP plane electric wave $E_A^i(r; d_A, p_A)$, $A = L, R$, incident upon the scatterer D_1. The scattering cross section σ_A^s expresses the scattered power

$$\sigma_A^s = \frac{2\sqrt{\mu}}{|p_A|^2 \sqrt{\varepsilon}} \langle P^s \rangle,$$

(4.35)

where

$$\langle P^s \rangle = \frac{1}{2} \text{Re} \int_S \hat{n} \cdot (E^s \times \overline{H^s}) ds$$

(4.36)

is the time-averaged scattered power. Taking into account that E^s, H^s satisfy (4.3)–(4.4), after some calculations, we find

$$\sigma_A^s = -\frac{\kappa}{\gamma^2 |p_A|^2} \text{Im} \int_S (\hat{n} \times E^s) \cdot (\nabla \times \overline{E^s} - \beta \gamma^2 \overline{E^s}) ds.$$

(4.37)

Let S_r be a sphere centered at the origin with radius r large enough to include the scatterer in its interior. Applying Gauss' theorem in the region between S and S_r, we get

$$\sigma_A^s = -\frac{\kappa}{\gamma^2 |\mathbf{p}_A|^2} \operatorname{Im} \int_{S_r} (\hat{\mathbf{n}} \times \mathbf{E}^s) \cdot (\nabla \times \overline{\mathbf{E}^s} - \beta \gamma^2 \overline{\mathbf{E}^s}) ds. \qquad (4.38)$$

For $r \to \infty$ we can use the asymptotic form and obtain

$$\sigma_A^s = \frac{1}{|\mathbf{p}_A|} \int_{S^2} \left[\frac{1}{\gamma_L^2} |\mathbf{g}_L(\hat{\mathbf{r}}; \hat{\mathbf{d}}_A, \mathbf{p}_A)|^2 + \frac{1}{\gamma_R^2} |\mathbf{g}_R(\hat{\mathbf{r}}; \hat{\mathbf{d}}_A, \mathbf{p}_A)|^2 \right] ds(\hat{\mathbf{r}}). \qquad (4.39)$$

The absorption cross section σ_A^a defines the total energy that is absorbed by a lossy scatterer and is given by

$$\sigma_A^a = \frac{2\sqrt{\mu}}{|\mathbf{p}_A|^2 \sqrt{\varepsilon}} \langle P^a \rangle, \qquad (4.40)$$

where

$$\langle P^a \rangle = -\frac{1}{2} \operatorname{Re} \int_S \hat{\mathbf{n}} \cdot (\mathbf{E}^t \times \overline{\mathbf{H}^t}) ds \qquad (4.41)$$

is the time-average absorbed power. As in the scattering cross section, we find

$$\sigma_A^a = \frac{\kappa}{\gamma^2 |\mathbf{p}_A|^2} \operatorname{Im} \int_{S_r} (\hat{\mathbf{n}} \times \mathbf{E}^t) \cdot (\nabla \times \overline{\mathbf{E}^t} - \beta \gamma^2 \overline{\mathbf{E}^t}) ds. \qquad (4.42)$$

The extinction cross section σ_A^e is given by

$$\sigma_A^e = \sigma_A^s + \sigma_A^a \qquad (4.43)$$

and describes the total energy that the scatterer extracts from the incident wave either by radiation or by absorption. In particular for the dielectric, inserting the transmission conditions into (4.42) and applying the divergence theorem in D_1 we take1 $\sigma_A^a = 0$.

Therefore, we can state the following optical theorem.

Theorem 4.6. *If $E_A^i(r; d_A, \mathbf{p}_A)$, $A = L, R$, is a plane electric wave incident upon a dielectric, then*

$$\sigma_A^s = -\frac{4\pi}{\gamma_A^2 |\mathbf{p}_A|^2} Re\{\overline{\mathbf{p}_A} \cdot \mathbf{g}_A(\hat{\mathbf{d}}_A; \hat{\mathbf{d}}_A, \mathbf{p}_A)\}. \qquad (4.44)$$

Proof. Apply general scattering theorem for $\hat{\mathbf{d}}_{A1} = \hat{\mathbf{d}}_{A2} = \hat{\mathbf{d}}_A$ and $\mathbf{p}_{A1} = \mathbf{p}_{A2} = \mathbf{p}_A$ and take into account the asymptotic expression of scattering cross section σ_A^s.

Next consider the following tangential subsets of $L^2(S^2)$:

$$T_L^2(S^2) = \{\mathbf{b}_L : S^2 \to \mathbb{C}^3 : \mathbf{b}_L \in L^2(S^2), \hat{\mathbf{n}} \cdot \mathbf{b}_L = 0, \hat{\mathbf{n}} \times \mathbf{b}_L = -i\mathbf{b}_L\} \qquad (4.45)$$

$$T_R^2(S^2) = \{\mathbf{b}_R : S^2 \to \mathbb{C}^3 : \mathbf{b}_R \in L^2(S^2), \widehat{\mathbf{n}} \cdot \mathbf{b}_R = 0, \widehat{\mathbf{n}} \times \mathbf{b}_R = i\mathbf{b}_R\} \qquad (4.46)$$

□

Definition 4.7. A LCP Beltrami Herglotz function is a function of the form

$$\mathbf{q}_L = \int_{S^2} \mathbf{b}_L(\widehat{\mathbf{d}}_L) e^{i\gamma\widehat{\mathbf{d}}_L \cdot \mathbf{r}} \mathrm{d}s(\widehat{\mathbf{d}}_L), \qquad (4.47)$$

where $\mathbf{b}_L \in T_L^2(S^2)$ is the LCP Beltrami Herglotz kernel. Similarly, an RCP Beltrami Herglotz function is a function of the form

$$\mathbf{q}_R = \int_{S^2} \mathbf{b}_R(\widehat{\mathbf{d}}_R) e^{i\gamma\widehat{\mathbf{d}}_R \cdot \mathbf{r}} \mathrm{d}s(\widehat{\mathbf{d}}_R), \qquad (4.48)$$

where $\mathbf{b}_R \in T_R^2(S^2)$ is respectively the RCP Beltrami Herglotz kernel.

Remark 4.8. It is easily seen that both LCP and RCP Beltrami Herglotz functions satisfy the Beltrami equations (4.6), are divergence free, and satisfy the vector Helmholtz equation

$$\nabla \times \nabla \times \mathbf{q}_A + \gamma_A^2 \mathbf{q}_A = \mathbf{0}, \, A = L, R \qquad (4.49)$$

Using the LCP and RCP Beltrami Herglotz functions we will introduce the concept of a chiral Herglotz pair

Definition 4.9. A chiral Herglotz pair is a pair of vector fields of the form

$$E(\mathbf{r}) = \int_{S^2} \mathbf{b}_L(\widehat{\mathbf{d}}_L) e^{i\gamma\widehat{\mathbf{d}}_L \cdot \mathbf{r}} \mathrm{d}s(\widehat{\mathbf{d}}_L) + \int_{S^2} \mathbf{b}_R(\widehat{\mathbf{d}}_R) e^{i\gamma\widehat{\mathbf{d}}_R \cdot \mathbf{r}} \mathrm{d}s(\widehat{\mathbf{d}}_R),$$

$$H(\mathbf{r}) = \frac{1}{i\eta} \left(\int_{S^2} \mathbf{b}_L(\widehat{\mathbf{d}}_L) e^{i\gamma\widehat{\mathbf{d}}_L \cdot \mathbf{r}} \mathrm{d}s(\widehat{\mathbf{d}}_L) - \int_{S^2} \mathbf{b}_R(\widehat{\mathbf{d}}_R) e^{i\gamma\widehat{\mathbf{d}}_R \cdot \mathbf{r}} \mathrm{d}s(\widehat{\mathbf{d}}_R) \right) \qquad (4.50)$$

that is

$$E(\mathbf{r}) = \mathbf{q}_L + \mathbf{q}_R,$$

$$H(\mathbf{r}) = \frac{1}{i\eta}(\mathbf{q}_L - \mathbf{q}_R). \qquad (4.51)$$

Theorem 4.10. *For given densities* $\mathbf{b}_A \in T_A^2(S^2)$, $A = L, R$, *the solution to the scattering problem for the incident wave*

$$E_{b_A}^i(\mathbf{r}) = \int_{S^2} b_A(\widehat{\mathbf{d}}_A) e^{i\gamma_A\widehat{\mathbf{d}}_A \cdot \mathbf{r}} \mathrm{d}s(\widehat{\mathbf{d}}_A) \qquad (4.52)$$

is given by the relation

$$E_{b_A}^s(\mathbf{r}) = \int_{S^2} \{E_L^s(\mathbf{r}; \widehat{\mathbf{d}}_A, b_A(\widehat{\mathbf{d}}_A)) + E_R^s(\mathbf{r}; \widehat{\mathbf{d}}_A, b_A(\widehat{\mathbf{d}}_A))\} \mathrm{d}s(\widehat{\mathbf{d}}_A) \qquad (4.53)$$

and has the far-field pattern

$$g(\hat{r})_{b_A} = \gamma_A \int_{S^2} \left[\frac{1}{\gamma_L} g_L(\hat{r}; \hat{d}_A, b_A(\hat{d}_A)) + \frac{1}{\gamma_R} g_R(\hat{r}; \hat{d}_A, b_A(\hat{d}_A)) \right] ds(d_A) \qquad (4.54)$$

Proof. We express the solutions of the scattering problem in terms of the boundary integral operators C, C_1, F, F_1 from second part, using (4.19) and we follow the analysis of ([19], pp. 188, lemma 6.31).

We will close this section by stating a new result relatively to general scattering theorem, where the incident field **E** is an electromagnetic chiral Herglotz incident field. In order to do this, we will define the LCP, RCP far-field operators. This result is important on solving the inverse scattering problem in chiral media.   □

Definition 4.11. The operators

$$F_A : T_A(S^2) \to T_A(S^2),$$

where $A = L, R$

$$(F_A b_A)(\hat{r}) := \gamma_A \int_{S^2} \left[\frac{1}{\gamma_L} g_L(\hat{r}; \hat{d}_A, b_A(\hat{d}_A)) + \frac{1}{\gamma_R} g_R(\hat{r}; \hat{d}_A, b_A(\hat{d}_A)) \right] ds(\hat{d}_A) \quad (4.55)$$

We state the following theorem [21] which gives the connection between the LCP, RCP far-field operator and the electromagnetic chiral LCP, RCP Herglotz pair, respectively.

Theorem 4.12. *Let $E^i_{b_A}$, $E^i_{b'_A}$, be electric chiral Herglotz functions with kernels b_A, $b'_A \in T_A(S^2)$, for $A = L, R$, and let E_{b_A}, $E_{b'_A}$ be solutions of the modified Helmholtz equation with $E^i_{b_A}$, $E^i_{b'_A}$ have replaced the incident field E^i. Then it holds,*

$$(F_A b'_A, b_A) + (b'_A, F_A b_A) = -\frac{\gamma_1}{2\pi} (F_A b'_A, F_A b_A), \text{ for } A = L, R. \qquad (4.56)$$

Proof. For the proof of this theorem, we refer to the proof of "general scattering theorem" Theorem 4.5 where the electric fields E_1, E_2 have been replaced by their superposition electric fields $E^i_{b_A}$ and $E^i_{b'_A}$, for $A = L, R$, respectively.   □

4.4   Low-Frequency Scattering

For the wave numbers γ_L, γ_R, γ_{1L}, γ_{1R}, from (4.7), we have

$$\gamma_L = \frac{\omega\sqrt{\varepsilon\mu}}{1 - \beta\omega\sqrt{\varepsilon\mu}}, \quad \gamma_R = \frac{\omega\sqrt{\varepsilon\mu}}{1 + \beta\omega\sqrt{\varepsilon\mu}}, \qquad (4.57)$$

$$\gamma_{1L} = \frac{\omega\sqrt{\varepsilon_1\mu_1}}{1 - \beta_1\omega\sqrt{\varepsilon_1\mu_1}}, \quad \gamma_{1R} = \frac{\omega\sqrt{\varepsilon_1\mu_1}}{1 + \beta_1\omega\sqrt{\varepsilon_1\mu_1}} \tag{4.58}$$

and as functions of the angular frequency ω, then they are analytic in a neighborhood of zero [10, 22], that is,

$$\gamma_L = \sum_{n=0}^{\infty} \frac{\omega^n}{n!}\gamma_L^{(n)}(0) = \sum_{n=0}^{\infty} \frac{\omega^n}{n!}\beta^{n-1}(\varepsilon\mu)^{\frac{n}{2}}n! = \sum_{n=0}^{\infty} \omega^n\beta^{n-1}(\varepsilon\mu)^{\frac{n}{2}} \tag{4.59}$$

$$\gamma_R = \sum_{n=0}^{\infty} \frac{\omega^n}{n!}\gamma_R^{(n)}(0) = \sum_{n=0}^{\infty} \frac{\omega^n}{n!}(-1)^{n+1}\beta^{n-1}(\varepsilon\mu)^{\frac{n}{2}}n! = \sum_{n=0}^{\infty} \omega^n(-1)^{n+1}\beta^{n-1}(\varepsilon\mu)^{\frac{n}{2}}$$
$$\tag{4.60}$$

Also, we obtain for the exponential terms

$$f_L(\omega) = e^{i\gamma_L\widehat{\mathbf{d}}_L\cdot\mathbf{r}}$$

$$= \sum_{n=0}^{\infty} \frac{\omega^n}{n!}f_L^{(n)}(0)$$

$$= i\widehat{\mathbf{d}}_L\cdot\mathbf{r}\sum_{n=0}^{\infty} \frac{\omega^n}{n!}\sum_{l=0}^{n-1}\binom{n-1}{l}\gamma_L^{n-l}(0)f_L^{(l)}(0)$$

$$= i\widehat{\mathbf{d}}_L\cdot\mathbf{r}\sum_{n=0}^{\infty} \frac{\omega^n}{n!}C_n, \tag{4.61}$$

$$f_R(\omega) = e^{i\gamma_R\widehat{\mathbf{d}}_R\cdot\mathbf{r}}$$

$$= \sum_{n=0}^{\infty} \frac{\omega^n}{n!}f_R^{(n)}(0)$$

$$= i\widehat{\mathbf{d}}_R\cdot\mathbf{r}\sum_{n=0}^{\infty} \frac{\omega^n}{n!}\sum_{l=0}^{n-1}\binom{n-1}{l}\gamma_R^{n-l}(0)f_R^{(l)}(0)$$

$$= i\widehat{\mathbf{d}}_R\cdot\mathbf{r}\sum_{n=0}^{\infty} \frac{\omega^n}{n!}D_n. \tag{4.62}$$

where $C_0 = D_0 = 1$. Thus, the incident field is analyzed as follows:

$$\mathbf{E}^{\text{inc}}(\mathbf{r}) = \sum_{n=0}^{\infty} \frac{\omega^n}{n!}\left[\mathbf{p}_L i(\widehat{\mathbf{d}}_L\cdot\mathbf{r})\sum_{l=0}^{n}C_n + \mathbf{p}_R i(\widehat{\mathbf{d}}_R\cdot\mathbf{r})\sum_{l=0}^{n}D_n\right] \tag{4.63}$$

The exterior field is also analyzed in two components, the exterior,

$$\mathbf{E} = \sum_{n=0}^{\infty} \frac{\omega^n}{n!}\varphi_n(\mathbf{r}) \tag{4.64}$$

and the interior,

$$\mathbf{E}_1 = \sum_{n=0}^{\infty} \frac{\omega^n}{n!} \varphi_n^1(\mathbf{r}) \tag{4.65}$$

Inserting the above expansions in equations and transmission conditions of (4.20) we obtain the following iterative sequence ($n \in \mathbb{N}$):

$$\nabla \times \nabla \times \varphi_0(\mathbf{r}) = \nabla \times \nabla \times \varphi_1(\mathbf{r}) = \mathbf{0} \ \mathbf{r} \in D,$$

$$\nabla \times \nabla \times \varphi_n(\mathbf{r}) = \varepsilon \mu n(n-1)[\beta^2 \nabla \times \nabla \times \varphi_{n-2}(\mathbf{r})$$

$$+ 2\beta \nabla \times \varphi_{n-2}(\mathbf{r}) + \varphi_{n-2}(\mathbf{r})] \ n \geqslant 2, \ \mathbf{r} \in D \tag{4.66}$$

$$\nabla \times \nabla \times \varphi_0^1(\mathbf{r}) = \nabla \times \nabla \times \varphi_1^1(\mathbf{r}) = \mathbf{0} \ \mathbf{r} \in D_1,$$

$$\nabla \times \nabla \times \varphi_n^1(\mathbf{r}) = \varepsilon_1 \mu_1 n(n-1)[\beta_1^2 \nabla \times \nabla \times \varphi_{n-2}^1(\mathbf{r})$$

$$+ 2\beta_1 \nabla \times \varphi_{n-2}^1(\mathbf{r}) + \varphi_{n-2}^1(\mathbf{r})] \ n \geqslant 2, \ \mathbf{r} \in D_1 \tag{4.67}$$

$$\widehat{\mathbf{n}} \times \varphi_n(\mathbf{r}) = \widehat{\mathbf{n}} \times \varphi_n^1(\mathbf{r}), \ n \geqslant 0, \ \mathbf{r} \in S, \tag{4.68}$$

$$\widehat{\mathbf{n}} \times \nabla \times \varphi_n^1(\mathbf{r}) - \varepsilon_1 \mu_1 \beta_1 n(n-1)[\beta_1 \widehat{\mathbf{n}} \times \nabla \times \varphi_{n-2}^1(\mathbf{r}) - \widehat{\mathbf{n}} \times \varphi_{n-2}^1(\mathbf{r})]$$

$$= \frac{\mu_1}{\mu} \widehat{\mathbf{n}} \times \nabla \times \varphi_n(\mathbf{r}) - \mu_1 \varepsilon \beta n(n-1)[\widehat{\mathbf{n}} \times \nabla \times \varphi_{n-2}(\mathbf{r}) - \widehat{\mathbf{n}} \times \varphi_{n-2}(\mathbf{r})] \ n \geqslant 2, \ \mathbf{r} \in S. \tag{4.69}$$

In order to derive radiation conditions for the low-frequency coefficients, we work as follows. First we construct an integral representation of the total exterior field in which all the data of the problem have been incorporated. In integral representation (4.27) applying Divergence Theorem to the interior of the scatterer we obtain

$$\int_S \widehat{\mathbf{n}} \cdot [\mathbf{E}(\mathbf{r}') \times \widetilde{B}(\mathbf{r}, \mathbf{r}')] ds(\mathbf{r}') = \int_{D_1} (\nabla \times \mathbf{E}) \cdot \widetilde{B}(\mathbf{r}, \mathbf{r}') du(\mathbf{r}')$$

$$- \int_{D_1} \mathbf{E}(\mathbf{r}') \cdot (\nabla_{\mathbf{r}} \times \widetilde{B}(\mathbf{r}, \mathbf{r}')) du(\mathbf{r}') \tag{4.70}$$

Taking into account the following properties of the dyadic Green function $[\widetilde{B}(\mathbf{r}, \mathbf{r}')]^{\top} = \widetilde{B}(\mathbf{r}', \mathbf{r})$ and $[\nabla_{\mathbf{r}} \times \widetilde{B}(\mathbf{r}, \mathbf{r}')]^{\top} = \nabla_{\mathbf{r}'} \times \widetilde{B}(\mathbf{r}, \mathbf{r}')$ and the vector-dyadic identities $\mathbf{E} \times \widetilde{B} = -[\widetilde{B}^{\top} \times \mathbf{E}]^{\top}$ and $\mathbf{E} \cdot \widetilde{B} = \widetilde{B}^{\top} \cdot \mathbf{E}$, we have

$$\int_S \widetilde{B}(\mathbf{r}, \mathbf{r}') \cdot (\widehat{\mathbf{n}} \times \mathbf{E}(\mathbf{r}')) ds(\mathbf{r}') = \int_{D_1} (\nabla \times \mathbf{E}_1) \cdot \widetilde{B}(\mathbf{r}', \mathbf{r}) du(\mathbf{r}')$$

$$- \int_S \mathbf{E}_1 \cdot (\nabla_{\mathbf{r}'} \times \widetilde{B}(\mathbf{r}', \mathbf{r})) ds(\mathbf{r}'), \tag{4.71}$$

and

$$\int_S \tilde{B}(\mathbf{r},\mathbf{r}') \cdot (\hat{\mathbf{n}} \times (\nabla \times \mathbf{E})(\mathbf{r}')) ds(\mathbf{r}') = 2\gamma_1^2 \beta_1 \int_{D_1} (\nabla \times \mathbf{E}_1(\mathbf{r}')) \cdot \tilde{B}(\mathbf{r}',\mathbf{r}) du(\mathbf{r}')$$

$$+\gamma_1^2 \int_{D_1} \mathbf{E}_1(\mathbf{r}') \cdot \tilde{B}(\mathbf{r}',\mathbf{r}) du(\mathbf{r}') - \int_{D_1} (\nabla \times \mathbf{E}_1(\mathbf{r}')) \cdot (\nabla_{\mathbf{r}'} \times \tilde{B}(\mathbf{r}',\mathbf{r})) du(\mathbf{r}') \quad (4.72)$$

and

$$\int_S \nabla_{\mathbf{r}} \times \tilde{B}(\mathbf{r},\mathbf{r}') \cdot (\hat{\mathbf{n}} \times \mathbf{E}(\mathbf{r}')) ds(\mathbf{r}') = \int_{D_1} (\nabla \times \mathbf{E}(\mathbf{r}')) \cdot (\nabla_{\mathbf{r}'} \times \tilde{B}(\mathbf{r}',\mathbf{r})) du(\mathbf{r}')$$

$$-2\beta\gamma^2 \int_{D_1} \mathbf{E}(\mathbf{r}') \cdot (\nabla_{\mathbf{r}'} \times \tilde{B}(\mathbf{r}',\mathbf{r})) du(\mathbf{r}') - \gamma^2 \int_{D_1} \mathbf{E}(\mathbf{r}') \cdot \tilde{B}(\mathbf{r}',\mathbf{r}) du(\mathbf{r}'). \quad (4.73)$$

Substituting the above three terms in (4.27) we finally get

$$\mathbf{E}(\mathbf{r}') = \mathbf{E}^{\text{inc}}(\mathbf{r}') + 2(\beta_1\gamma_1^2 - \beta\gamma^2) \int_{D_1} \nabla \times \mathbf{E}_1(\mathbf{r}') \cdot \tilde{B}(\mathbf{r}',\mathbf{r}) du(\mathbf{r}')$$

$$+(\gamma_1^2 - \gamma^2) \int_{D_1} \mathbf{E}_1(\mathbf{r}') \cdot \tilde{B}(\mathbf{r}',\mathbf{r}) du(\mathbf{r}') \quad (4.74)$$

Next we express all the fields in terms of low-frequency coefficients. The asymptotic representation for the nth order coefficient can be derived from this integral representation if we omit the nth term which is of order $\frac{1}{r}$. For convenience and to avoid long calculations we describe the above procedure for the zeroth-order approximation. We will give the solution of the problem in ellipsoidal geometry.

$$\nabla \times \nabla \times \varphi_0(\mathbf{r}) = \mathbf{0} \text{ in } D,$$

$$\nabla \times \nabla \times \varphi_0^1(\mathbf{r}) = \mathbf{0} \text{ in } D_1,$$

$$\nabla \cdot \varphi_0(\mathbf{r}) = \mathbf{0} \text{ in } D,$$

$$\nabla \cdot \varphi_0^1(\mathbf{r}) = \mathbf{0} \text{ in } D_1,$$

$$\hat{\mathbf{n}} \times \varphi_0(\mathbf{r}) = \hat{\mathbf{n}} \times \varphi_0^1(\mathbf{r}) \text{ on } S, \quad (4.75)$$

$$\hat{\mathbf{n}} \times (\nabla \times \varphi_0(\mathbf{r})) = \frac{\mu}{\mu_1}\hat{\mathbf{n}} \times \nabla \times \varphi_0^1(\mathbf{r}) \text{ on } S,$$

$$\varphi_0(\mathbf{r}) = \hat{\mathbf{q}}_L + \hat{\mathbf{q}}_R + O\left(\frac{1}{r}\right) \quad r \to \infty$$

$$\varphi_0(\mathbf{r}) = \mathbf{P}_0(\mathbf{r}) + \mathbf{W}_0(\mathbf{r}), \quad (4.76)$$

where $\mathbf{P}_0(\mathbf{r}) = \hat{\mathbf{q}}_L + \hat{\mathbf{q}}_R$ and $\mathbf{W}_0(\mathbf{r}) = \nabla U_0(\mathbf{r})$, if considering

$$U_0(\mathbf{r}) = \alpha_{00} I_0(\rho) + \alpha_{01}^1 F^1(\rho,\mu,\nu) + \alpha_{01}^2 F^2(\rho,\mu,\nu) + \alpha_{01}^3 F^3(\rho,\mu,\nu).$$

Similarly, for the interior approximation we obtain

$$\varphi_0^1(\mathbf{r}) = \mathbf{W}_0^1(\mathbf{r}) = \nabla U_0^1(\mathbf{r}) \tag{4.77}$$

where $U_0^1(\mathbf{r}) = c_{01}^1 \mathbf{E}_1^1(\rho,\mu,\nu) + c_{01}^2 \mathbf{E}_1^2(\rho,\mu,\nu) + c_{01}^3 \mathbf{E}_1^3(\rho,\mu,\nu)$.

We also write $\mathbf{d} = (d_1, d_2, d_3) = \widehat{\mathbf{q}}_L + \widehat{\mathbf{q}}_R$. Hence, we have for the exterior and interior zeroth approximation

$$\varphi_0(\mathbf{r}) = (d_1, d_2, d_3) + \big(A_1(\rho), A_2(\rho), A_3(\rho)\big)$$

$$-\frac{d_1(\varepsilon - \varepsilon_1)\alpha_1\alpha_2\alpha_3}{h_2 h_3[\varepsilon - (\varepsilon - \varepsilon_1)(1 - \alpha_1\alpha_2\alpha_3 I^1(\alpha_1))]} \cdot \frac{\mu\nu}{\rho\sqrt{\rho^2 - \mu^2}\sqrt{\rho^2 - \nu^2}}\widehat{\rho}$$

$$-\frac{d_2(\varepsilon - \varepsilon_1)\alpha_1\alpha_2\alpha_3}{h_3 h_1[\varepsilon - (\varepsilon - \varepsilon_1)(1 - \alpha_1\alpha_2\alpha_3 I^2(\alpha_1))]} \cdot \frac{\sqrt{\mu^2 - h_3^2}\sqrt{h_3^2 - \nu^2}}{\sqrt{\rho^2 - h_3^2}\sqrt{\rho^2 - \mu^2}\sqrt{\rho^2 - \nu^2}}\widehat{\rho}$$

$$-\frac{d_3(\varepsilon - \varepsilon_1)\alpha_1\alpha_2\alpha_3}{h_1 h_2[\varepsilon - (\varepsilon - \varepsilon_1)(1 - \alpha_1\alpha_2\alpha_3 I^3(\alpha_1))]} \cdot \frac{\sqrt{\mu^2 - h_3^2}\sqrt{h_3^2 - \nu^2}}{\sqrt{\rho^2 - h_2^2}\sqrt{\rho^2 - \mu^2}\sqrt{\rho^2 - \nu^2}}\widehat{\rho}$$

$$\tag{4.78}$$

where

$$A_1(\rho) = \frac{d_1(\varepsilon - \varepsilon_1)\alpha_1\alpha_2\alpha_3 I^1(\rho)}{\varepsilon - (\varepsilon - \varepsilon_1)\alpha_1\alpha_2\alpha_3 I^1(\alpha_1)},$$

$$A_2(\rho) = \frac{d_2(\varepsilon - \varepsilon_1)\alpha_1\alpha_2\alpha_3 I^2(\rho)}{\varepsilon - (\varepsilon - \varepsilon_1)\alpha_1\alpha_2\alpha_3 I^2(\alpha_1)},$$

$$A_3(\rho) = \frac{d_3(\varepsilon - \varepsilon_1)\alpha_1\alpha_2\alpha_3 I^3(\rho)}{\varepsilon - (\varepsilon - \varepsilon_1)\alpha_1\alpha_2\alpha_3 I^3(\alpha_1)}$$

$$\varphi_0^1(\mathbf{r}) = (B_1, B_2, B_3) \tag{4.79}$$

where

$$B_1 = \frac{d_1\varepsilon}{\varepsilon - (\varepsilon - \varepsilon_1)\alpha_1\alpha_2\alpha_3 I^1(\alpha_1)},$$

$$B_2 = \frac{d_2\varepsilon}{\varepsilon - (\varepsilon - \varepsilon_1)\alpha_1\alpha_2\alpha_3 I^2(\alpha_1)},$$

$$B_3 = \frac{d_3\varepsilon}{\varepsilon - (\varepsilon - \varepsilon_1)\alpha_1\alpha_2\alpha_3 I^3(\alpha_1)}).$$

Appendix

Ellipsoidal Harmonics

The interior ellipsoidal harmonics are given by the Lamé products

$$\mathbb{E}_n^m(\rho,\mu,\nu) = \mathbf{E}_n^m(\rho)\mathbf{E}_n^m(\mu)\mathbf{E}_n^m(\nu) \tag{4.80}$$

and the exterior ellipsoidal harmonics

$$\mathbb{F}_n^m(\rho,\mu,\nu) = \mathbb{F}_n^m(\rho)\mathbf{E}_n^m(\mu)\mathbf{E}_n^m(\nu) \tag{4.81}$$

where $\mathbb{E}_n^m$, $\mathbb{F}_n^m$ are the Lamé functions of the first and second kind, respectively. The Lamé functions are related by the formula

$$\mathbb{F}_n^m(\rho) = (2n+1)\mathbf{E}_n^m(\rho)\mathbf{I}_n^m(\rho), \tag{4.82}$$

where the functions $\mathbf{I}_n^m(\rho)$ are the elliptic integrals

$$\mathbf{I}_n^m(\rho) = \int_\rho^{+\infty} \frac{du}{[\mathbf{E}_n^m(u)]^2\sqrt{u^2 - h_2^2}\sqrt{u^2 - h_3^2}} \tag{4.83}$$

The value of the function $\mathbf{I}_n^m(\rho)$ on the surface $\rho = \alpha_1^j$ of the m.l.e. is denoted by $\mathbf{I}_n^{(j)m}$. The index n specifies the degree of the corresponding ellipsoidal harmonic and it takes the values $0,1,2,\ldots$. The superscript m represents the number of independent harmonic functions of degree n and runs through the values $1,2,\ldots,2n+1$. In this work we only use the ellipsoidal harmonics of degree $0,1$, which are given in their Cartesian form

$$\mathbb{E}_0^m(\rho,\mu,\nu) = 1\mathbb{E}_1^m(\rho,\mu,\nu) = \frac{h_1h_2h_3}{h_m}x_m, \quad m = 1,2,3. \tag{4.84}$$

The exterior ellipsoidal harmonics $\mathbb{F}_n^m(\rho,\mu,\nu)$ of degree $0,1$ are given by (4.83), when (4.82) and (4.84) are used. The Lamé functions $\mathbf{E}_n^m$ of degree $0,1$ that appear in the expressions (4.83) for the elliptic integrals $\mathbf{I}(\rho)$ are:

$$\mathbf{E}_0^1(\rho) = 1$$

$$\mathbf{E}_1^m(\rho) = \sqrt{\rho^2 - (\alpha_1^j)^2 + (\alpha_1^m)^2}, \quad m = 1,2,3.$$

The outward normal derivative on the surface $\rho = \alpha_1^j$ is given by

$$\partial_n = \frac{\alpha_2^j\alpha_3^j}{\sqrt{(\alpha_1^j)^2 - \mu^2}\sqrt{(\alpha_1^j)^2 - \nu^2}}\partial_\rho. \tag{4.85}$$

For details we refer to Hobson's book [23].

References

1. Ammari, H., Nedelec, J.C.: Time-harmonic electromagnetic fields in chiral media. Modern Methods Diff. Theory Appl. Eng. **42**, 174–202 (1997)
2. Ammari, H., Laoudi, M., Nedelec. J.: Low frequency behavior of solutions to electromagnetic scattering problems in chiral media. SIAM J. Appl. Math. **58**, 1022–1042 (1998)
3. Ammari, H., Hamdache, K., Nedelec, J.C.: Chirality in the Maxwell equations by the dipole approximation method. SIAM J. Appl. Math. **59**, 2045–2059 (1999)
4. Allam, A.M.RI.: Chiral absorbing material. In: Seventeenth National Radio Science Conference, Minufga University , Egypt, 2000
5. Arago, D.F.J.: Sur une modification remarquable qu'eprouvent les rayons lumineux dans leur passage a travers certains corps diaphanes, et sur quelques autres nouveaux phenomenes d'optique., Mem. Inst. 12, Part I, 93–134 (1811)
6. Athanasiadis, C.: Low-frequency electromagnetic scattering for a multi-layered scatterer. Quart. Jl. Mech. Appl. Math. **44**, 55–67 (1991)
7. Athanasiadis, C., Martin, P., Stratis, I.G.: Electromagnetic scattering by a homogeneous chiral obstacle: boundary integral relations and low-chirality approximations. SIAM J. Appl. Math. **59**, 1745–1762 (1999)
8. Athanasiadis, C., Costakis, G., Stratis, I.G.: Electromagnetic scattering by a homogeneous chiral obstacle in a chiral environment. IMA J. Appl. Math. **64**, 245–258 (2000)
9. Athanasiadis, C., Costakis, G., Stratis, I.G.: Electromagnetic scattering by a perfectly conducting obstacle in a homogeneous chiral environment: solvability and low frequency theory. Math. Meth. Appl. Sci. **25**, 927–944 (2002)
10. Athanasiadis, C., Costakis, G., Stratis, I.G.: Transmission problems in contrasting chiral media. Rep. Math. Phys. **53**, 143–156 (2004)
11. Athanasiadis, C.: On the far field patterns for electromagnetic scattering by a chiral obstacle in a chiral environment. Math.Anal. Appl. **309**, 517–533 (2005)
12. Athanasiadis, C., Kardasi, E.: Beltrami Herglotz functions for electromagnetic scattering in chiral media. Appl. Anal. **84**, 1–19 (2005)
13. Athanasiadis, C., Kardasi E.: On the far field operator for electromagnetic scattering in chiral media. Applicable Analysis. **85**, 623–639 (2006)
14. Athanasiadis, C., Berketis, N.: Scattering relations for point-source excitation in chiral media. Math. Meth. Appl. Sci. **29**, 27–48 (2006)
15. Bohren, C.F.: Light scattering by an oprtically active sphere. Chem. Phys. Lett. **29**, 458–462 (1974)
16. Bohren, C.F.: Scattering of an electromagnetic wave by an optically active cylinder. J. Colloid Interface Sci. **66**, 105–109 (1978)
17. Biot, J.B.: Memorire sur le rotations que certains substances impriment aux axes de polarization de rayons lumineux. Memoiries de la classe de sciences Mathematiques et Physiques de l'Institut Imperial de France II, 41–136 (1812)
18. Colton, D., Kress, R.: Integral Equation Methods in Scattering Theory. Wiley, New York (1983)
19. Colton, D., Kress R.: Inverse Acoustic and Electromagnetic Scattering Theory. Springer, Berlin (1992)
20. Colton, D., Kress, R.: Eigenvalues of the far field operator and inverse scattering theory. SIAM J. Math. Anal. **26**(3), 601–615 (1995)
21. Colton, D., Kress, R.: Eigenvalues of the far field operator for the Helmholtz equation in an absorbing medium. SIAM J. Appl. Math. **26**(6), 1724–1735 (1995)
22. Dassios, G., Kleinman, R.: Low Frequency Scattering Theory. Clarendon Press, Oxford (2000)
23. Hobson, E.W.: The Theory of Spherical and Ellipsoidal Harmonics. Chelsea, New York (1955)
24. Lakhtakia, A., Varadan, V., Varadan K.: Time-harmonic electromagnetic fields in chiral media. Lecture Notes in Physics, No 335, Springer, Berlin (1989)

25. Lakhtakia, A., Varadan, V., Varadan, K.: Effective properties of a sparse random distribution of non-interacting small chiral spheres in a chiral host medium. J. Phys. D. Appl. Phys. **24**, 1–6 (1991)
26. Lakhtakia, A.: Beltrami Fields in Chiral Media. World Scientific, Singapore (1994)
27. Lindell, I.V., Sihvola, A.H., Tretyakov, S.A., Viitanen A. J.: Electromagnetic waves in chiral and Bi-isotropic media. Artech House, Boston (1994)
28. Ola, P.: Boundary integral equations for the scattering of electromagnetic waves by a homogeneous chiral obstacle. J.Math.Phys. **35**, 3969–3980 (1994)
29. Pasteur, L.C.R.: Acad. Sci. Paris, **26**, 48–49 (1848)
30. Pasteur, L.C.R.: Acad. Sci. Paris, **28**, 56–59 (1850)
31. Roach, G.F.: Green's Functions. Cambridge University Press, London (1982)

Chapter 5
Orthonormality in Interpolation Schemes for Reconstructing Signals

Nicholas J. Daras

Abstract Given only a few of initial Fourier coefficients for a continuous-time periodic signal, we construct efficient rational approximants to the whole signal everywhere on his domain of definition. The convergence of these approximants depends on the orthonormality of the chosen generating polynomial system $\{V_{m+1}(e^{it}) : m = 0, 1, \dots\}$ into $L^2[-\pi, \pi]$. The form of each $V_{m+1}(x)$ is characterized by recurrence relations due to the connection between Schur and Szegö theories.

Keywords Approximation by rational functions • Padé-type and/or Padé approximants • Fourier coefficients • Orthogonal polynomials • Acceleration of convergence • Interpolation • Trigonometric approximation and interpolation

Mathematics Subject Classification (2010): 41A20, 41A21, 42A16, 42C05, 65B99, 65D05, 65T40

5.1 Introduction

Given a signal, say a sound or a image, Fourier analysis easily calculates the frequencies and the amplitudes of those frequencies which make up the signal. However, Fourier methods are not always a good tool to recapture the signal, particularly if it is highly non-smooth: too much Fourier information is needed to reconstruct the signal locally. Especially, there is no analytical way to reconstruct with exactitude a noised signal if only a few of its initial Fourier coefficients are known.

N.J. Daras (✉)
Department of Military Science, Hellenic Army Academy,
166 73, Vari Attikis, Greece
e-mail: darasn@sse.gr

N.J. Daras (ed.), *Applications of Mathematics and Informatics in Military Science*,
Springer Optimization and Its Applications 71, DOI 10.1007/978-1-4614-4109-0_5,
© Springer Science+Business Media New York 2012

Such extension problems in Taylor series context have a long history dating from 1907 [6,7]. Indeed, the Schur problem, or Carathéodory–Fejér problem, was to find conditions for the existence of an analytic function bounded by one in the unit planar disk whose initial Taylor coefficients are given numbers $a_0, a_1, \ldots, a_\mu$. In [25, 26], Schur showed that such a function exists if and only if the lower triangular matrix

$$(a_{\mu,k} = a_{\mu-k})_{k \leq \mu} = \begin{pmatrix} a_0 & 0 & 0 & 0 \\ a_1 & a_0 & 0 & 0 \\ & \ddots & \ddots & 0 \\ a_\mu & & a_1 & a_0 \end{pmatrix}$$

is bounded by one as an operator on complex Euclidean space, and he determined how all solutions can be found. The method was adapted to Pick–Nevanlinna interpolation by Nevanlinna [24]. Ever since, several different applications and analogue developments are considered. For instance, given a partial covariance sequence of length μ, the problem of finding all positive rational extensions of degree at most μ is a fundamental open problem with important applications in signal processing and speech processing [5, 13, 15, 18–21, 23] and in stochastic realization theory and system identification [2, 22, 25].

In this paper, we will consider *a numerical version of the above Carathéodory–Schur interpolation problem in the trigonometric context*. In particular, *we will investigate a numerical method for constructing efficient approximants to any continuous-time periodic signal by using only a few of its initial Fourier coefficients*. These approximants are real parts of rational functions with numerators determined by the condition that the Fourier series expansion of the approximants matches the Fourier series of the signal as far as possible. Motivated by this crucial property, the obtained approximants will be called Padé-type and/or Padé approximants. The convergence of these approximants depends strongly on the orthonormality of the chosen generating polynomial system $\{V_{m+1}(e^{it}) : m = 0, 1, \ldots\}$ into $L^2[-\pi, \pi]$. The form of $V_{m+1}(x)$ is characterized by recurrence relations dues to the connection between Schur and Szegö theories.

The detailed definition of a Padé-type and a Padé approximant to a continuous-time periodic L^1-signal and their principal properties are presented in Sect. 5.1. Section 5.2 is devoted to a study of the convergence of a sequence of such approximants. Section 5.3 deals with assumptions under which, for every sequence of functions converging to a periodic continuous signal there is a Padé-type approximation sequence converging pointwise to that signal faster than the first sequence. Finally, in Sect. 5.4 numerical examples are given making use of Padé-type approximants.

5.2 Construction of Rational Approximants to a Periodic Continuous-Time Signal

Consider a T-periodic continuous-time signal $\varphi(x)$ defined over an interval $[-(T/2),(T/2)]$. Suppose $\varphi \in L^1(-(T/2),(T/2))$ and f has a finite number of extrema and discontinuities in any given interval. Then, φ has a Fourier series expansion defined by $\varphi(x) = \sum_{v=-\infty}^{\infty} \sigma_v e^{iv(2\pi/T)x}$ with $\sigma_v = \frac{1}{T}\int_{-(T/2)}^{(T/2)} \varphi(y)e^{-iv(2\pi/T)y}dy$ ($v = 0,\pm 1,\pm 2,\ldots$). The problem we will investigate is the following: "if only a few Fourier coefficients $\sigma_0,\sigma_{\pm 1},\ldots,\sigma_{\pm \mu}$ of the signal $\varphi(x)$ are given, construct efficient rational approximants to the whole signal $\varphi(x)$ (almost) everywhere on $[-(T/2),(T/2)]$".

Putting $t := (2\pi/T)x$, the signal $\varphi(x)$ converts to a 2π-period continuous-time signal $f(t)$ defined over the standard interval $[-\pi,\pi]$, with Fourier series expansion $f(t) = \sum_{v=-\infty}^{\infty} c_v e^{ivt}$ and Fourier coefficients $c_v = \frac{1}{2\pi}\int_{-\pi}^{\pi} f(s)e^{-ivs}ds$ ($v = 0,\pm 1,\pm 2,\ldots$). It is clear that $c_v = \sigma_v$ ($v = 0,\pm 1,\pm 2,\ldots$) and f can be identified with a real-valued function $u(z)$ in L^1 of the unit circle C by setting $u(e^{it}) := f(t) = \sum_{v=-\infty}^{\infty} c_v e^{ivt}$. From the solution of the Dirichlet problem in the unit disk D, it follows that the extended real-valued function $u(z) := u(re^{it})$ is harmonic in the open unit disk and such that $\lim_{r \to 1} \|u_r(t) - u(e^{it})\|_1 (:= \lim_{r \to 1}\int_{-\pi}^{\pi} \lceil u_r(t) - u(e^{it})\rceil dt) = 0$ and the Fourier series expansion of the restriction $u_r(t)$ of $u(re^{it})$ to any circle C_r of radius $r < 1$ is given by $\sum_{v=-\infty}^{\infty} c_v r^{|v|}e^{ivt}$.

We can consider Padé-type and Padé approximants to the harmonic function $u(z)$. These approximants can be chosen in such a way to be harmonic real-valued functions everywhere in D. Their fundamental property is that the Fourier series expansion of their restriction to any circle of radius $r < 1$ coincides with the Fourier series expansion of $u(z)$ as far as possible. Indeed, $u(z)$ is the real part of an analytic function $F(z)$ in D. So, $u = F + \overline{F}$ where $\overline{F}$ denotes complex conjugate of F. If $\sum_{v=0}^{\infty} a_v z^v$ is the Taylor power series expansion of F around $0 \in D$, then $u(z) = 2\text{Re}(F(z)) - a_0 = 2\text{Re}(\sum_{v=0}^{\infty} a_v z^v) - a_0$ and $a_v = c_v$ ($v = 0,1,\ldots$). Define the $\mathbb{C}$-linear functional $T_f : \mathbb{P}(\mathbb{C}) \to \mathbb{C}; x^v \mapsto T_f(x^v) := a_v(= c_v)$, where $\mathbb{P}(\mathbb{C})$ is the vector space of all complex analytic polynomials. An application of Cauchy's integral formula shows that $|T_f(p(x))| \leq (2\pi)^{-1}\sup_{|s|=r}|F(s)|\sup_{|s|=r^{-1}}|p(s)|$ whenever $p(x) \in \mathbb{P}(\mathbb{C})$. By density, there is a continuous extension of T_f into the space $\mathcal{O}(\overline{D})$ of all functions which are analytic in an open neighborhood of $\overline{D}$. In particular, for every fixed point $z \in D$, the number $T_f((1 - xz)^{-1})$ is well defined and equals $\sum_{v=0}^{\infty} a_v z^v$. Hence, it holds $u(z) = 2\text{Re}T_f((1 - xz)^{-1}) - c_0$ for any $z \in D$. If the function $(1 - xz)^{-1}$ is replaced by a polynomial $Q(x,z)$, then $u(z)$ is approximated by $2\text{Re}T_f(Q(x,z)) - c_0$. This is an approximate quadrature formula and leads to a Padé or, more generally, to a Padé-type approximant to the harmonic real-valued function $u(z)$.

Definition 5.1. [10, 11]

(i) Let $Q_m(x,z)$ denote the unique complex polynomial of degree at most m in x, which interpolates the Cauchy kernel $(1-xz)^{-1}$ at $m+1$ points $\pi_0, \pi_1, \ldots, \pi_m$, i.e., $Q_m(\pi_k, z) = (1-\pi_k z)^{-1}$ for any $z \in \mathbb{C} \setminus \{\pi_k^{-1}; \ k = 0, 1, \ldots, m\}$ and $k \leq m$. The real-valued function $(m/m+1)_u(z) := 2\mathrm{Re}T_f(Q_m(x,z)) - c_0$ is said to be a Padé-type approximant to $u(z)$, with generating polynomial $V_{m+1}(x) = \gamma \prod_{k=0}^{m}(x - \pi_k)$ $(\gamma \in \mathbb{C} \setminus \{0\})$.

(ii) Suppose the Hankel determinants

$$H_{m+1}^{(f)}(c_0) = \det \underbrace{\begin{pmatrix} c_0 & c_1 & c_2 & \cdots & c_m \\ c_1 & c_2 & c_3 & \cdots & c_{m+1} \\ c_2 & c_3 & c_4 & \cdots & c_{m+2} \\ \vdots & \vdots & \vdots & & \vdots \\ c_m & c_{m+1} & c_{m+2} & \cdots & c_{2m} \end{pmatrix}}_{m+1}$$

are all different from zero. There exists uniquely defined family $\{q_{m+1}(x) : m = 0, 1, 2, \ldots\}$ of orthogonal polynomials with respect to the functional T_f, in the sense that $T_f(x^\nu q_{m+1}(x)) = 0$ for any $\nu = 0, 1, 2, \ldots, m$. The exact degree of each $q_{m+1}(x)$ is $m+1$. Let $\mathcal{P}_m(x,z)$ denote the unique complex polynomial of degree at most m in x, which interpolates the Cauchy kernel $(1-xz)^{-1}$ at the $m+1$ zeros $\rho_0, \rho_1, \ldots, \rho_m$ of $q_{m+1}(x)$, i.e., $\mathcal{P}_m(\pi_k, z) = (1-\rho_k z)^{-1}$ for any $z \in \mathbb{C} \setminus \{\rho_k^{-1}; \ k = 0, 1, \ldots, m\}$ and $k \leq m$. The real-valued function $[m/m+1]_u(z) := 2\mathrm{Re}T_f(\mathcal{P}_m(x,z)) - c_0$ is said to be a Padé approximant to $u(z)$.

Obviously, the Padé approximant $[m/m+1]_u(z)$ to $u(z)$, if it exists, can be viewed as a Padé-type approximant with predesigned generating polynomial $V_{m+1}(x) = q_{m+1}(x)$.

The approximants in Definition 5.1 can be interpreted as real parts of rational functions. Indeed, we put

$$\widetilde{V}_{m+1}(z) := z^{m+1} V_{m+1}(x^{-1}), \qquad \widetilde{q}_{m+1}(z) := z^{m+1} q_{m+1}(z^{-1}),$$
$$W_m(z) := T_f([V_{m+1}(x) - V_{m+1}(z)]/[x-z]), \ w_m(z) := T_f([q_{m+1}(x) - q_{m+1}(z)]/[x-z]),$$
$$\widetilde{W}_m(z) := z^m W_m(z^{-1}), \qquad \widetilde{w}_m(z) := z^m w_m(z^{-1}).$$

As it is well known, the general Hermite interpolation polynomial can be deduced from the Lagrange polynomial by continuity arguments. So, by using the expression of the Lagrange interpolation polynomial for $(1 - xz)^{-1}$, one can exploit the definitions of $W_m(z)$ and $w_m(z)$ to create appropriate partial fraction decompositions and thus, to obtain the following.

Theorem 5.2. [10, 11]

(i) $(m/m+1)_u(z)$ and $[m/m+1]_u(z)$ are real parts of rational complex functions of type $(m, m+1)$:

$$(m/m+1)_u(z)=2Re\left(\frac{\widetilde{W}_m(z)}{\widetilde{V}_{m+1}(z)}\right)-c_0 \text{ and } [m/m+1]_u(z)=2Re\left(\frac{\widetilde{w}_m(z)}{\widetilde{q}_{m+1}(z)}\right)-c_0.$$

(ii) The errors of the respective approximations equal

$$(m/m+1)_u(z) - u(z) = 2Re\left[\frac{1}{V_{m+1}(z^{-1})}T_f\left(\frac{V_{m+1}(x)}{xz-1}\right)\right] \text{ and }$$

$$[m/m+1]_u(z) - u(z) = 2Re\left[\frac{1}{[q_{m+1}]^2(z^{-1})}T_f\left(\frac{[q_{m+1}]^2(x)}{xz-1}\right)\right]$$

$$= Re\left[\frac{1}{[q_{m+1}]^2(z^{-1})}T_f\left(\frac{x^{m+1}q_{m+1}(x)}{xz-1}\right)\right].$$

Now, from the exactitude of the Newton–Côtes quadrature formula for polynomials of degree less than m, it follows that the Fourier series expansion of the restriction $(m/m+1)_{u_r}(t)$ of $(m/m+1)_u(-e^{it})$ to any circle of radius $r < 1$ matches the Fourier series expansion of the restriction $u_r(t)$ of $u(z)$ to that circle up to the $\pm m$th Fourier term. This property justifies the notation Padé-type approximant to $u(z)$. Similarly, from the exactitude of the Gauss quadrature formula for polynomials of degree less than $2m + 1$, it follows that the Fourier series expansion of the restriction $[m/m+1]_{u_r}(t)$ of $[m/m+1]_u(re^{it})$ to any circle of radius $r < 1$ matches the Fourier series expansion of the restriction $u_r(t)$ of $u(z)$ to that circle up to the $\pm(2m+1)$th Fourier term. This property justifies the notation Padé approximant to $u(z)$. Summarizing, we have the next formulation for the crucial property of Padé-type and Padé approximation.

Theorem 5.3. The fundamental property of these approximants is the following.

- The Fourier series expansion $\sum_{\nu=-\infty}^{\infty}d_\nu^{(m)}r^{|\nu|}e^{i\nu t}$ of the restriction $(m/m+1)_{u_r}(t)$ of $(m/m+1)_u(re^{it})$ to the circle of radius r fulfills $d_\nu^{(m)} = c_\nu$ for any $\nu = 0, \pm1, \pm2, \ldots, \pm m$.
- The Fourier series expansion $\sum_{\nu=-\infty}^{\infty}f_\nu^{(m)}r^{|\nu|}e^{i\nu t}$ of the restriction $[m/m+1]_{u_r}(t)$ of $[m/m+1]_u(re^{it})$ to the circle of radius r fulfills $f_\nu^{(m)} = c_\nu$ for any $\nu = 0, \pm1, \pm2, \ldots, \pm(2m)$.

Since the two limits $\lim_{r \to 1}(m/m+1)_{u_r}(t)$ and $\lim_{r \to 1}[m/m+1]_{u_r}(t)$ are uniform on $[-\pi, \pi]$, the two functions $(m/m+1)_u(re^{it})$ and $[m/m+1]_u(re^{it})$ are the Poisson integrals of two continuous functions on the unit circle.

Definition 5.4. [10, 11]

(i) The radial limit $(m/m+1)_f(t) := \lim_{r \to 1}(m/m+1)_{u_r}(t) = 2\mathrm{Re}T_f(Q_m(x,e^{it})) - c_0$ is said to be a Padé-type approximant to $f(t)$, with generating polynomial $V_{m+1}(x) = \gamma\prod_{k=0}^m(x - \pi_k)$ $(\gamma \in \mathbb{C} \setminus \{0\})$.

(ii) The radial limit $[m/m+1]_f(t) := \lim_{r \to 1}[m/m+1]_{u_r}(t) = 2\mathrm{Re}T_f(\mathcal{P}_m(x,e^{it})) - c_0$ is said to be a Padé approximant to $f(t)$.

Remark 5.5. From a numerical point of view, construction of a Padé-type approximant $(m/m+1)_f(t)$ with generating polynomial $V_{m+1}(x) = \gamma\prod_{k=0}^m(x - \pi_k)$ needs the knowledge of the $(2m+1)$ first Fourier coefficients $c_0, c_{\pm 1}, \ldots, c_{\pm m}$, while construction of the Padé approximant $[m/m+1]_f(t)$ makes use of the $(4m+1)$ Fourier coefficients $c_0, c_{\pm 1}, \ldots, c_{\pm 2m}$.

By Theorems 5.2 and 5.3, we immediately get the following fundamental properties of these approximants.

Theorem 5.6. *(i) The Padé-type approximant $(m/m+1)_f(t)$ is the real part of a rational complex function of type $(m, m+1)$: $(m/1)_f(t) = 2\mathrm{Re}(\widetilde{W}_m(e^{it})/\widetilde{V}_{m+1}(e^{it})) - c_0$. The error of the respective approximation is given by the following theoretical formula*

$$(m/m+1)_f(t) = \frac{1}{\pi}\lim_{r \to 1} Re\left\{\frac{1}{V_{m+1}(r^{-1}e^{-it})}\int_{-\pi}^\pi \frac{f(s)V_{m+1}(e^{-is})}{re^{i(t-s)}-1}ds\right\}$$

where the limit is taken in the L^1-norm. The Fourier series expansion $\sum_{v=-\infty}^\infty d_v^{(m)}\, e^{ivt}$ of $(m/m+1)_f(t)$ fulfills $d_v^{(m)} = c_v$ for any $v = 0, \pm 1, \pm 2, \ldots, \pm m$.

(ii) *The Padé approximant $[m/m+1]_f(t)$ is also the real part of a rational complex function of type $(m, m+1)$: $[m/m+1]_f(t) = 2\mathrm{Re}(\widetilde{w}_m(e^{it})/\widetilde{q}_{m+1}(e^{it})) - c_0$. The error of the respective approximation equals*

$$[m/m+1]_f(t) - f(t) = \frac{1}{\pi}\lim_{r \to 1} Re\left\{\frac{1}{[q_{m+1}]^2(r^{-1}e^{-it})}\int_{-\pi}^\pi \frac{f(s)[q_{m+1}]^2(e^{-is})}{re^{i(t-s)}-1}ds\right\}$$

where the limit is taken in the L^1-norm. The Fourier series expansion $\sum_{v=-\infty}^\infty \beta_v^{(m)}\, e^{ivt}$ of $[m/m+1]_f(t)$ fulfills $\beta_v^{(m)} = c_v$ for any $v = 0, \pm 1, \pm 2, \ldots, \pm(2m+1)$.

Proof. To prove the theorem, we will use the standard identification of the 2π-periodic signal $f(t)$ $(t \in [-\pi, \pi])$ with the L^1-function $u(z) \equiv u(e^{it})$ $(z = e^{it}, |z| = 1)$.

(i) Let $(r_n \in [0,1])_{n=0,1,2,\ldots}$ be any sequence such that $\lim_{n \to +\infty} r_n = 1$. Since, by Theorem 5.2.(i), $(m/m+1)_u(r_n e^{it}) = 2\mathrm{Re}(\widetilde{W}_m(r_n e^{it})/\widetilde{V}_{m+1}(r_n e^{it})) - c_0$, the uniform convergence of the sequence $((m/m+1)_u(r_n e^{it}))_{n=0,1,2,\ldots}$ to the radial limit function $(m/m+1)_u(e^{it})$ implies that $(m/m+1)_u(e^{it}) = 2\mathrm{Re}(\widetilde{W}_m(e^{it})/\widetilde{V}_{m+1}(e^{it})) - c_0$, and the first assertion of part (i) is proved. To prove the second assertion, recall that $\lim_{n \to +\infty} \|u(r_n e^{it}) - u(e^{it})\|_1 = 0$. Further, by the uniform convergence of the sequence $((m/m+1)_u(r_n e^{it}))_{n=0,1,2,\ldots}$

to the radial limit function $(m/m+1)_u(e^{it})$ we also get $\lim_{n \to +\infty} \|(m/m+1)_u(r_n e^{it}) - (m/m+1)_u(e^{it})\|_1 = 0$. Letting $\varepsilon > 0$, we infer that there exists a $N = N(\varepsilon)$ such that $\|u(r_n e^{it}) - u(e^{it})\|_1 + \|(m/m+1)_u(r_n e^{it}) - (m/m+1)_u(e^{it})\|_1 < \varepsilon$ whenever $n \geq N$, and therefore $\|[(m/m+1)_u(e^{it}) - u(e^{it})] - [(m/m+1)_u(r_n e^{it}) - u(r_n e^{it})]\|_1 < \varepsilon$ for any $n \geq N$, or, by Theorem 5.2.(ii),

$$\left\| [(m/m+1)_u(e^{it}) - u(e^{it})] - 2\mathrm{Re}\left[\frac{1}{V_{m+1}(r_n^{-1}e^{-it})} T_f\left(\frac{V_{m+1}(x)}{xr_n e^{it} - 1} \right) \right] \right\|_1 < \varepsilon$$

for any $n \geq N$. Now, set $V_{m+1}(x) = \sum_{k=0}^{m+1} b_k^{(m)} x^k$. An application of the continuity property for the linear functional T_f shows that

$$T_f\left(\frac{V_{m+1}(x)}{xr_n e^{it} - 1} \right) = -\sum_{v=0}^{\infty} r^v e^{ivt} \sum_{k=0}^{m} \frac{b_k^{(m)}}{2\pi} \int_{-\pi}^{\pi} u(e^{is}) e^{-i(v+k)s} ds.$$

Computing, we obtain

$$T_f\left(\frac{V_{m+1}(x)}{xr_n e^{it} - 1} \right) = \frac{1}{2\pi} \int_{-\pi}^{\pi} u(e^{is}) \left[-\left(\sum_{k=0}^{m} b_k^{(m)} e^{-iks} \right)\left(\sum_{v=0}^{\infty} r_n^v e^{i(t-s)v} \right) \right]$$

$$= \frac{1}{2\pi} \int_{-\pi}^{\pi} u(e^{is}) \left[\frac{V_{m+1}(e^{-is})}{r_n e^{i(t-s)} - 1} \right] ds = \int_{-\pi}^{\pi} \frac{f(s)V_{m+1}(e^{-is})}{r_n e^{i(t-s)} - 1} ds.$$

Hence

$$\left\| [(m/m+1)_u(e^{it}) - u(e^{it})] - 2\mathrm{Re}\left[\frac{1}{V_{m+1}(r_n^{-1}e^{-it})} \int_{-\pi}^{\pi} \frac{f(s)V_{m+1}(e^{-is})}{r_n e^{i(t-s)} - 1} ds \right] \right\|_1 < \varepsilon,$$

for any $n \geq N$. Using the standard identifications $(m/m+1)_u(e^{it}) = (m/m+1)_f(t)$ and $u(e^{it}) = f(t)$, we conclude that

$$\left\| [(m/m+1)_f(t) - f(t)] - 2\mathrm{Re}\left[\frac{1}{V_{m+1}(r_n^{-1}e^{-it})} \int_{-\pi}^{\pi} \frac{f(s)V_{m+1}(e^{-is})}{r_n e^{i(t-s)} - 1} ds \right] \right\|_1 < \varepsilon,$$

for any $n \geq N$, which proves the second assertion of part (i). It remains to prove the third assertion of part (i). Since every harmonic real-valued function in the unit disk, with continuous boundary values, is the Poisson integral of its continuous restriction to the unit circle, we have

$$(m/m+1)_u(re^{it}) = \frac{1}{2\pi} \int_{-\pi}^{\pi} (m/m+1)_u(e^{i\theta}) \frac{1-r^2}{1 - 2r\cos(t-\theta) + r^2} d\theta$$

$$= \frac{1}{2\pi} \int_{-\pi}^{\pi} (m/m+1)_u(e^{i\theta}) \sum_{v=-\infty}^{\infty} r^{|v|} e^{iv(t-\theta)} d\theta$$

$$= \sum_{v=-\infty}^{\infty} r^{|v|} \left[\frac{1}{2\pi} \int_{-\pi}^{\pi} (m/m+1)_u(e^{i\theta}) e^{iv\theta} d\theta \right] e^{it}.$$

Henceforth, the Fourier series expansion of $(m/m+1)_u(\mathrm{re}^{it})$ is $\sum_{v=+\infty}^{\infty} d_v^{(m)}$ $r^{|v|}e^{iv+t}$. From Theorem 5.3, it follows that $d_v^{(m)} = c_v$ for any $v = 0, \pm 1, \pm 2, \ldots,$ $\pm m$, and the proof of part (i) is complete.

(ii) Repetition of the arguments in (i) with only obvious formal changes proves (ii).

$\square$

From Theorem 5.6, it follows immediately that if only a few Fourier coefficients c_v of the signal $f(t)$ are given, then one can approximate f by an approximant in the Padé-type and/or Padé sense: $(m/m+1)_f(t) \approx f(t)$ and $[m/m+1]_f(t) \approx f(t)$.

Remark 5.7. If $f : [-\pi, \pi] \to \mathbb{R} : t \mapsto f(t)$ is a 2π-periodic continuous-time signal with Fourier series representation $\sum_{v=-\infty}^{\infty} c_v e^{ivt}$, the function $(m+n/m+1)_f(t) :=$ $2\mathrm{Re}\left(\sum_{v=-\infty}^{n-1} c_v e^{ivt} + e^{int} T_{f_n}(Q_m(x,e^{it}))\right) - c_0$ $(t \in [-\pi, \pi])$ is the real part of a rational function of type $(m+n, m+1)$ with respect to the dependent variable $s = s(t) = e^{it}$ and is said to be a *Padé-type approximant of type* $(m+n, m+1)$ *to the signal* f. The functional T_{f_n} is now defined by $T_{f_n}(x^v) = c_{n+v}$. The fundamental property of such an approximant is the coincidence of its Fourier representation with that of f up to the $\pm(m+n)$th Fourier term. Similarly, the function $[m+n/m+$ $1]_f(t) := 2\mathrm{Re}\left(\sum_{v=-\infty}^{n-1} c_v e^{ivt} + e^{int} T_{f_n}(\mathcal{P}_m(x,e^{it}))\right) - c_0$ $(t \in [-\pi, \pi])$ is the real part of a rational function of type $(m+n, m+1)$ with respect to the dependent variable $s = s(t) = e^{it}$ and is said to be a *Padé approximant of type* $(m+n, m+1)$ *to the signal* f. The fundamental property of such an approximant is the coincidence of its Fourier representation with that of f up to the $\pm(2m+n)$th Fourier term.

5.3 Convergence and Orthogonal Polynomials

In this section, we shall study the convergence of a sequence of Padé-type approximants to a signal. The corresponding results derived for a sequence of Padé approximants can be viewed as a very special case. Let $\mathcal{M} = (\pi_{m,k})_{m=0,1,\ldots;k=0,1,\ldots,m}$ be an infinite triangular interpolation matrix with complex entries $\pi_{m,k} \in D$. For any fixed $z \in \mathbb{C} \setminus \{\pi_{m,k}^{-1}; m = 0,1,\ldots$ and $k = 0,1,\ldots,m\}$, let $Q_m(x,z)$ denote the unique polynomial of degree at most m in x, which interpolates the Cauchy kernel $(1 - xz)^{-1}$ in the $(m+1)$ nodes of the mth row of $\mathcal{M}$, i.e., $Q_m(\pi_{m,k}, z) = (1 - \pi_{m,k}z)^{-1}$ for any $k \leq m$.

Theorem 5.8. *Suppose the family* $\{V_{m+1}(e^{it}) : m = 0,1,\ldots\}$ *of generating polynomials is an orthonormal bounded system in* $L^2[-\pi, \pi]$ *and* $\inf_{m \geq M_0} \inf_{t \in [-\pi, \pi]}$ $|V_{m+1}(e^{it})| > 0$, *for a* M_0. *Then for any real-valued* 2π-*periodic continuous-time signal* $f(t) \in L^1[-\pi, \pi]$, *the associated sequence* $((m/m + 1)_f(t) =$ $2\mathrm{Re} T_f(Q_m(x,e^{it})) - c_0)_{m=0,1,\ldots}$ *of Padeé-type approximants to* $f(t)$ *with generating polynomials* $V_{m+1}(x) = \prod_{k=0}^{m}(x - \pi_{m,k})$ *converges to* $f(t)$ *almost everywhere on* $[-\pi, \pi]$. *Especially, if* $f(t) \in C[-\pi, \pi]$, *the sequence* $((m/m + 1)_f(t) =$ $2\mathrm{Re}(\widetilde{W}_m(e^{it})/\widetilde{V}_{m+1}(e^{it})) - c_0)_{m=0,1,\ldots}$ *converges to* $f(t)$ *everywhere on* $[-\pi, \pi]$.

Proof. Let $f(t) \in L^1[-\pi, \pi]$ be a real-valued 2π-periodic continuous-time signal. Let also $\varepsilon > 0$ and $(r_n)_{n=0,1,2,\ldots}$ be a strictly increasing sequence of positive numbers such that $\lim_{n \to \infty} r_n = 1$. Fix any n. By Theorem 5.6.(i), there is a subsequence $(r_{n_j})_{j=0,1,2,\ldots}$ of (r_n) such that

$$(m/m+1)_f(t) - f(t) = \frac{1}{\pi} \lim_{j \to \infty} \mathrm{Re}\left\{ \frac{1}{V_{m+1}(r_{n_j}^{-1} e^{-it})} \int_{-\pi}^{\pi} f(s) \frac{V_{m+1}(e^{-is})}{(r_{n_j} e^{i(t-s)} - 1)} ds \right\}$$

for almost all $t \in [-\pi, \pi]$. Denote by $\mathcal{D}$ the set of all points $t \in [-\pi, \pi]$ with this property. For every $t \in \mathcal{D}$, one can find a $J = J(\varepsilon.m)$ such that

- $|(m, m+1)_f(t) - f(t)| \leq \frac{1}{\pi}\left|\frac{1}{V_{m+1}(r_{n_j}^{-1} e^{-it})}\right| \left| \int_{-\pi}^{\pi} f(s)\left[\frac{V_{m+1}(e^{-is})}{(r_{n_j} e^{i(t-s)} - 1)}\right] ds\right| + \frac{\varepsilon}{2} \ (j \geq J)$

- The function $[-\pi - \pi] \to \mathbb{C} : s \mapsto (f(s)/[r_{n_j} e^{i(t-s)} - 1])$ is in $L^1[-\pi, \pi]$, so, by Mercer's Theorem, the Fourier coefficients of this function with respect to the orthonormal family $\{V_{m+1}(e^{-is}) : m = 0, 1, \ldots\}$ tend to zero. This means that there exists a $M = M(\varepsilon, t)$ such that

$$\pi^{-1}\left|V_{m+1}(r_{n_j}^{-1} e^{-it})^{-1}\right| \left|\int_{-\pi}^{\pi}[f(s)V_{m+1}(e^{-is}/(r_{n_j} e^{i(t-s)} - 1)]ds\right| < \varepsilon/2 \ (m \geq M \text{ and } j \geq J).$$

Combining of these inequalities shows that $|(m/m+1)_f(t) - f(t)| < \varepsilon$ for any $m \geq M$. This implies that $\lim_{m \to \infty}(m/m+1)_f(t) = f(t)$ for almost all $t \in [-\pi, \pi]$. If $f(t) \in C[-\pi, \pi]$, the convergence holds for any $t \in [-\pi, \pi]$. $\square$

Corollary 5.9. *Let the generating polynomials of a Padé-type approximation* $V_{m+1}(x) = \sum_{k=0}^{m+1} b_k^{(m)} x^k = \prod_{k=0}^m (x - \pi_{m,k})$ $(m = 0, 1, \ldots)$ *be such that* $\sum_{k=0}^{m+1} |b_k^{(m)}|^2 = (1/2\pi)$ $(m \geq 0)$ *and* $\sum_{k=0}^{m+1} b_k^{(m)} \overline{b_k^{(n)}} = 0$ $(n \geq m)$. *If there are two constants* $\sigma < \infty$ *and* $\tau < 1$ *fulfilling* $\sum_{k=0}^{m+1} |b_k^{(m)}| < \sigma$ $(m \geq 0)$ *and* $|\pi_{m,k}| < \tau$ $(m \geq 0 \text{ and } 0 \leq k \leq m)$, *then, for any real-valued 2π-periodic continuous-time signal* $f(t) \in L^1[-\pi, \pi]$, *the associated sequence* $((m/m+1)_f(t) = 2\mathrm{Re}[\widetilde{W}_m(e^{it}/\widetilde{V}_{m+1}(e^{it})] - c_0)_{m=0,1,\ldots}$ *of Padé-type approximants to* $f(t)$, *with generating polynomials* $V_{m+1}(x) = \prod_{k=0}^m (x - \pi_{m,k})$, *converges to* $f(t)$ *almost everywhere on* $[-\pi, \pi]$. *Especially, if* $f(t) \in C[-\pi, \pi]$, *the sequence* $((m/m+1)_f(t) = 2\mathrm{Re}(\widetilde{W}_m(e^{it})/\widetilde{V}_{m+1}(e^{it})) - c_0)_{m=0,1,\ldots}$ *converges to* $f(t)$ *everywhere* $[-\pi, \pi]$.

Proof. It is easily seen that if there is a constant $\tau < 1$ satisfying $|\pi_{m,k}| < \tau$ $(m \geq 0 \text{ and } 0 \leq k \leq m)$, then there exists an open neighborhood U of the unit circle into which there holds $\inf_{z \in U} |V_{m+1}(z)| \geq K > 0$ for some positive constant K $(m = 0, 1, 2, \ldots)$ which is independent of m. In particular, we have $\inf_{m \geq M_0} \inf_{t \in [-\pi, \pi]} |V_{m+1}(e^{it})| > 0$ for a M_0. Further, if the polynomial $V_{m+1}(x) = \prod_{k=0}^m (x - \pi_{m,k})$ is written in the form $V_{m+1}(x) = \sum_{k=0}^{m+1} b_k^{(m)} x^k$, the orthonormality assumption for the family $\{V_{m+1}(e^{it}) : m = 0, 1, \ldots\}$ is completely described by the following conditions: $\sum_{k=0}^{m+1} |b_k^{(m)}|^2 = (1/2\pi)$ $(m \geq 0)$ and $\sum_{k=0}^{m+1} b_k^{(m)} \overline{b_k^{(n)}} = 0$ $(n \geq m)$. In fact, for any $m \geq 0$, we have

$$\bullet \; 2\pi \sum_{k=0}^{m+1} |b_k^{(m)}|^2 = \sum_{k=0}^{m+1} b_k^{(m)} \overline{b_v^{(m)}} \int_{-\pi}^{\pi} e^{iks} e^{-ivs} ds = \int_{-\pi}^{\pi} V_{m+1}(e^{is}) \overline{V_{m+1}(e^{is})} ds$$

$$= \int_{-\pi}^{\pi} |V_{m+1}(e^{is})|^2 ds = 1$$

$$\bullet \; 2\pi \sum_{k=0}^{m+1} b_k^{(m)} \overline{b_k^{(n)}} = \sum_{k=0}^{m+1} \sum_{v=0}^{n+1} b_k^{(m)} \overline{b_v^{(n)}} \int_{-\pi}^{\pi} e^{iks} e^{-ivs} ds$$

$$= \int_{-\pi}^{\pi} V_{m+1}(e^{is}) \overline{V_{n+1}(e^{is})} ds = 0 \; (n > m).$$

Finally, the boundedness assumption for the family $\{V_{m+1}(e^{it}) : m = 0, 1, \ldots\}$ is guaranteed if there is a positive constant $\sigma < \infty$ satisfying $\sum_{k=0}^{m+1} |b_k^{(m)}| < \sigma$ for any m. ∎

Remark 5.10. In [12], we gave a stronger sufficient condition in terms of the entries $\pi_{m,k}$ only: *if the interpolation points $\pi_{m,k}$ are chosen so that $-1 < \pi_{m,k} < 1$ and $\lim_{m \to \infty} \sum_{n>1} \frac{1}{n} \sum_{k=0}^{m} (\pi_{m,k})^n = -\infty$, then for any real-valued 2π-periodic continuous-time signal $f(t) \in C[-\pi, \pi]$, the associated sequence $((m/m+1)_f(t))_{m=0,1,\ldots}$ of Padé-type approximants to $f(t)$, with generating polynomials $V_{m+1}(x) = \prod_{k=0}^{m} (x - \pi_{m,k})$, converges to $f(t)$ everywhere on $[-\pi, \pi]$.* If, for instance,

$$\pi_{m,k} = \begin{cases} 0, & \text{if } k = \text{even} \\ -a, & \text{if } k = \text{odd} \end{cases}$$

for some $a \in [0, 1]$, then $\lim_{m \to \infty} (m/m+1)_f(t) = f(t)$ whenever $t \in [-\pi, \pi]$.

Remark 5.11. Another approach to the convergence problem is analogous to the results of Eiermann [14]: *if the generating polynomials $V_{m+1}(x)$ of a Padé-type approximation satisfy $\lim_{m \to \infty} (V_{m+1}(x)/V_{m+1}(z^{-1})) = 0$ uniformly on the compact subsets of an open set $\Omega \subset \mathbb{C}^2$ containing $(\overline{D} \times D) \cup (\mathbb{C} \times \{0\})$, then, for any real-valued 2π-periodic continuous-time signal $f(t) \in C[-\pi, \pi]$, the associated sequence $((m/m+1)_f(t))_{m=0,1,\ldots}$ of Padé-type approximants to $f(t)$, with generating polynomials $V_{m+1}(x) = \prod_{k=0}^{m} (x - \pi_{m,k})$, converges to $f(t)$ everywhere on $[-\pi, \pi]$* [9, 10].

As it is mentioned in Theorem 5.8, the crucial property for the convergence of a sequence of Padé-type approximants to a signal is the orthonormality of the system $\{V_{m+1}(e^{it}) : m = 0, 1, \ldots\}$ into $L^2[-\pi, \pi]$. One can obtain interesting results about the form of $V_{m+1}(x)$, dues to the connection between Schur and Szegö theories. This connection is often attributed to Akhiezer [1], but it appears earlier and in greater detail in the papers of Geronimo [16, 17]. It is based on important recurrence relations which were first given in previous Szegö's work [28].

Theorem 5.12. [16, 17] *Denoting by $V_{m+1}^*(x)$ the polynomial $x^{m+1} \overline{V_{m+1}(1/\bar{x})}$, the recurrence relations written in terms of the monic polynomials $V_{m+1}(x) = V_{m+1}(x)/V_{m+1}^*(0)$ are of the general form $V_{m+2}(x) = x V_{m+1}(x) - \bar{a}_{m+1} V_{m+1}^*(x)$ for certain parameters $a_{m+1} \in \mathbb{C}$ $(m = 0, 1, \ldots)$.*

Remark 5.13. In current terminology the numbers $-\bar{a}_{m+1} = V_{m+2}(0)$ are called *Szegö parameters.*

Let us now see what the Schur parameters are. To do so, we first remind the following Schur's construction.

Theorem 5.14. *There is a one-to-one correspondence between the class $S(D)$ of analytic functions which are bounded by one on the unit disk D and the set of all sequences $(\gamma_{m+1})_{m=0,1,2,...}$ of complex numbers which are bounded by 1 and such that if some term has unit modulus, then all subsequent terms are zero.*

Proof. Given any $\phi(x) \in S(D)$, define a sequence

$$\phi_{m+1}(x) = \begin{cases} \phi(x), & \text{if } m = 0 \\ \frac{\phi_m(x) - \phi_m(0)}{x[1 - \overline{\phi_m(0)}\phi_m(x)]}, & \text{if } m \geq 1. \end{cases}$$

If $|\phi_i(0)| = 1$ for some i, then $\phi_i(x)$ is constant and we take $\phi_j(x) = 0$ for any $j > i$. This occurs if and only if $\phi(x)$ is finite Blaschke product of i factors:

$$\phi(x) = c\frac{x - b_1}{1 - \overline{b_1}x}\frac{x - b_2}{1 - \overline{b_2}x} \cdots \frac{x - b_i}{1 - \overline{b_i}x},$$

where $|c| = 1$ and $b_1, b_2, \ldots, b_i$ are points in D. The numbers $\gamma_{m+1} = \gamma_{m+1}^{(\phi)} := \phi_{m+1}(0)$ $(m \geq 0)$ are defined to be the *Schur parameters for $\phi(x)$.*   □

Remark 5.15. The method of labeling $S(D)$ by numerical sequences is known as the Schur algorithm and is due to Schur [26, 27].

Since $g(x) = (2\pi)^{-1} \int_{-\pi}^{\pi}([e^{is} + x]/[e^{is} - x])ds$ has positive real part in the open unit disk D and value 1 at the origin, the function $\Phi(x) = (1/x)([g(x) - 1]/[g(x) + 1])$ belongs to $S(D)$. In 1943, Geronimo showed that

Theorem 5.16. [17] *The Schur parameters $\gamma_{m+1} = \gamma_{m+1}^{(\Phi)}$ for $\Phi(x)$ coincide with the numbers a_{m+1} in the recurrence formula of Theorem 5.12: $\gamma_{m+1} = \gamma_{m+1}^{(\Phi)} = a_{m+1}$ $(m = 0, 1, \ldots)$.*

5.4 Acceleration of the Convergence and Orthogonal Polynomials

We shall now see how Padé-type approximants can accelerate the convergence of functional sequences. More precisely, we shall study assumptions under which, for every sequence of functions converging to a real-valued continuous 2π-periodic signal, there is always a Padé-type approximation sequence converging pointwise to that signal faster than the first sequence. This property, due to the free choice of the interpolation points $\pi_{m,k}$ permits us to construct better and better approximations to continuous functions.

Using techniques similar those proposed by Bromwich [4] and Clark [8], we can prove the following.

Proposition 5.17. *Let Δ be the operator of differences. Let also $(x_m)_{m=0,1,2,...}$ and $(y_m)_{m=0,1,2,...}$ be two sequences of real numbers satisfying $x_m \neq y_m$ and $\lim_{m\to\infty} y_m = 0$. Suppose $(y_m)_{m=0,1,2,...}$ is strictly monotone. If $\lim_{m\to\infty}[\Delta x_m / \Delta y_m] = 0$, then there is a $x \in \mathbb{R}$ such that $\lim_{m\to\infty} x_m = x$ and $\lim_{m\to\infty}[(x_m - x)/y_m] = 0$. In other words, under the assumptions of Proposition 5.17, the sequence $(x_m)_{m=0,1,2,...}$ converges to x faster than the sequence $(y_m)_{m=0,1,2,...}$ converges to 0.*

Proof. Let $\varepsilon < 0$. Without loss of generality, we can assume that the sequence $(y_m)_{m=0,1,2,...}$ is strictly decreasing. Since $\lim_{m\to\infty}[\Delta x_m/\Delta y_m] = 0$, there is a $M_0 > 0$ so that $-\varepsilon < \Delta x_m/\Delta y_m < \varepsilon$ for any $m \geq M_0$. Since $\Delta y_m < 0$, these inequalities can be rewritten as $-\varepsilon(y_m - y_{m+1}) < x_m - x_{m+1} < \varepsilon(y_m - y_{m+1})$, for any $m \geq M_0$. Replacing the index m by $m+1, m+2, \ldots, m+p-1$, we form the p inequalities $-\varepsilon(t_{m+j-1} - y_{m+j}) < x_{m+j-1} - x_{m+j} < \varepsilon(y_{m+j-1} - y_{m+j})$ $(j = 1, 2, \ldots, p)$ for any $m \geq M_0$. Adding these inequalities, we get $-\varepsilon(y_m - y_{m+p}) < x_m - x_{m+p} < \varepsilon(y_m - y_{m+p})$ for any $m \geq M_0$. Of course, we can suppose $y_m - y_{m+p} < 1$ and therefore obtain $-\varepsilon < x_m - x_{m+p} < \varepsilon$ for any $m \geq M_0$. It follows that $(x_m)_{m=0,1,2,...}$ is a Cauchy sequence. As the real field $\mathbb{R}$ is complete, this sequence converges to a limit in $\mathbb{R}$, say x. Letting now $p \to \infty$ in the inequalities $-\varepsilon(y_m - y_{m+p}) < x_m - x_{m+p} < \varepsilon(y_m - y_{m+p})$ $(m \geq M_0)$, we take $-\varepsilon y_m < x_m - x < \varepsilon y_m$ for any $m \geq M_0$, which implies that $\lim_{m\to\infty}[(x_m - x)/y_m] = 0$. The proof is now complete. $\qquad\square$

Corollary 5.18. *Let Δ be the operator of differences. Let also $(x_m)_{m=0,1,2,...}$ and $(y_m)_{m=0,1,...}$ be two convergent sequences of real numbers. Suppose $(y_m)_{m=0,1,2,...}$ is strictly monotone and $\lim_{m\to\infty} y_m = 0$. If*

$$\overline{\lim}_{m\to\infty} |\Delta x_m|^{1/m} = r < R = \lim_{m\to\infty} |\Delta y_m|^{1/m} = 0,$$

the sequence $(x_m)_{m=0,1,2,...}$ converges faster than the sequence $(y_m)_{m=0,1,2,...}$.

Proof. It is well known that if $\overline{\lim}_{m\to\infty} |\Delta x_m|^{1/m} = r < R = \lim_{m\to\infty} |\Delta y_m|^{1/m} = 0$, the sequence $(\Delta x_m)_{m=0,1,2,...}$ converges faster than the sequence $(\Delta y_m)_{m=0,1,2,...}$ [3]. Hence, from Proposition 5.17, it follows immediately the desired assertion. $\qquad\square$

Combination of Corollary 5.18 with Theorem 5.8 gives the following convergence acceleration result.

Theorem 5.19. *Let $(y_m)_{m=0,1,2,...}$ be any strictly monotone converging sequence. Suppose the family $\{V_{m+1}(e^{it}) : m = 0, 1, \ldots\}$ is an orthonormal bounded system in $L^2[-\pi, \pi]$ and $\inf_{m \geq M_0} \inf_{t \in [-\pi,\pi]} |V_{m+1}(e^{it})| > 0$, for a M_0. Then, for any real-valued 2π-periodic continuous-time signal $f(t) \in C[-\pi, \pi]$, the associated* sequence $\left((m/m+1)_f(t) = 2\mathrm{Re}\left(\frac{\widetilde{W}_m(e^{it})}{\widetilde{V}_{m+1}(e^{it})}\right)\right)_{m=0,1,\dots}$ of Padé-type approximants converges to $f(t)$ faster than $(y_m)_{m=0,1,2,\dots}$ everywhere on

$$\left\{t \in [-\pi, \pi] : \right.$$

$$\lim_{m \to \infty} |\Delta y_m|^{\frac{1}{m}} \geq \overline{\lim}_{m \to \infty} \left[\sup_{|x| \leq 1} \left|\frac{V_{m+1}(x)V_{m+2}(e^{-it}) - V_{m+1}(e^{-it})V_{m+2}(x)}{1 - xe^{it}}\right|\right]^{\frac{1}{m}}\right\}.$$

Proof. Obviously, condition $\inf_{m \geq M_0} \inf_{t \in [-\pi,\pi]} |V_{m+1}(e^{it})| > 0$ (for a M_0) implies that there is a positive constant K and a neighborhood U of the unit circle into which the generating polynomials $V_{m+1}(x)$ satisfy $|V_{m+1}(x)| \geq K$ for any $x \in U$ and m sufficiently large. By Theorem 5.8 the sequence $((m/m+1)_f(t))_{m=0,1,\dots}$ converges to the signal $f(t)$ everywhere on $[-\pi, \pi]$. Letting $t \in [-\pi, \pi]$ be fixed, it is easily seen that

$$|\Delta(m/m+1)_f(t)|^{1/m} \leq 2|T_f(Q_{m+1}(x, e^{it}) - Q_m(x, e^{it}))|$$

where $Q_m(x,z)$ denotes the unique complex polynomial of degree at most m in x, which interpolates the Cauchy kernel $(1 - xz)^{-1}$ at $m+1$ points $\pi_0, \pi_1, \dots, \pi_m$, i.e., $Q_m(\pi_k, z) = (1 - \pi_k z)^{-1}$ for any $z \in \mathbb{C} \setminus \{\pi_k^{-1}; k = 0, 1, \dots, m\}$ and $k \leq m$. Thus, the continuity of the linear functional T_f implies that

$$|\Delta(m/m+1)_f(t)|^{1/m}$$

$$\leq \left(\frac{\mathfrak{L}_f}{K}\right)^{1/m} \left[\sup_{|x| \leq 1} \left|\frac{V_{m+1}(x)V_{m+2}(e^{-it}) - V_{m+1}(e^{-it})V_{m+2}(x)}{1 - xe^{it}}\right|\right]^{1/m},$$

where the constant $\mathfrak{L}_f$ depends only on f. By passing in the upper limit, we obtain

$$\overline{\lim}_{m \to \infty} |\Delta(m/m+1)_f(t)|^{\frac{1}{m}}$$

$$\leq \overline{\lim}_{m \to \infty} \left[\sup_{|x| \leq 1} \left|\frac{V_{m+1}(x)V_{m+2}(e^{-it}) - V_{m+1}(e^{-it})V_{m+2}(x)}{1 - xe^{it}}\right|\right]^{\frac{1}{m}}.$$

Application of Corollary 5.18 for the sequences $(x_m = (m/m+1)_f(t))_{m=0,1,2,\dots}$ and $(y_m)_{m=0,1,2,\dots}$ proves the Theorem. $\qquad \square$

Similarly, using Corollary 5.9 instead of Theorem 5.8, we are leaded to the

Theorem 5.20. *Let the generating polynomials of a Padé-type approximation* $V_{m+1}(x) = \sum_{k=0}^{m+1} b_k^{(m)} x^k = \prod_{k=0}^{m} (x - \pi_{m,k})$ $(m = 0, 1, \dots)$ *be such that* $\sum_{k=0}^{m+1} |b_k^{(m)}|^2 = (1/2\pi)$ $(m \geq 0)$ *and* $\sum_{k=0}^{m+1} b_k^{(m)} \overline{b_k^{(n)}} = 0$ $(n \geq m)$. *Assume that there are two constants* $\sigma < \infty$ *and* $\tau < 1$ *fulfilling* $\sum_{k=0}^{m+1} |b_k^{(m)}| < \sigma$ $(m \geq 0)$ *and* $|\pi_{m,k}| < \tau$ $(m \geq 0$ *and*

$0 \leq k \leq m$). *Then, for any real-valued 2π-periodic continuous-time signal $f(t) \in C[-\pi, \pi]$, the associated sequence $((m/m + 1)_f(t) = 2Re(\widetilde{W}_m(e^{it})/\widetilde{V}_{m+1}(e^{it})) - c_0)_{m=0,1,...}$ of Padé-type approximants converges to $f(t)$ everywhere on $[-\pi, \pi]$ faster than every strictly monotone converging sequence $(y_m)_{m=0,1,2,...}$ satisfying*

$$\lim_{m \to \infty} |\Delta y_m|^{1/m} > \overline{\lim}_{m \to \infty} \left[\sum_{k=0}^{m+2} \sum_{\substack{v=0 \\ (v \neq k)}}^{m+1} |b_k^{(m+1)} b_v^{(m)}| \right]^{1/m}.$$

Proof. We note that

$$|V_{m+1}(x)V_{m+2}(e^{-it}) - V_{m+1}(e^{-it})V_{m+2}(x)|$$

$$= \left| \sum_{v=0}^{m+1} b_v^{(m)} e^{-vt} \sum_{k=0}^{m+2} b_k^{(m+1)} x^k - \sum_{k=0}^{m+2} b_k^{(m+1)} e^{-ikt} \sum_{v=0}^{m+1} b_v^{(m)} x^v \right|$$

$$= \left| \sum_{k=0}^{m+2} \sum_{v=0}^{m+1} b_k^{(m+1)} b_v^{(m)} [x^k e^{-vt} - x^v e^{-ikt}] \right|$$

$$\leq \sum_{k=0}^{m+2} \sum_{v=0}^{m+1} |b_k^{(m+1)} b_v^{(m)}| |x^k e^{-vt} - x^v e^{-ikt}|.$$

Thus

$$\overline{\lim}_{m \to \infty} \left[\sup_{|x| \leq 1} \left| \frac{V_{m+1}(x)V_{m+2}(e^{-it}) - V_{m+1}(e^{-it})V_{m+2}(x)}{1 - xe^{it}} \right| \right]^{1/m}$$

$$\leq \overline{\lim}_{m \to \infty} \left[\sum_{k=0}^{m+2} \sum_{\substack{v=0 \\ (v \neq k)}}^{m+1} |b_k^{(m+1)} b_v^{(m)}| \right]^{1/m}.$$

Application of Theorem 5.19 completes the proof.   □

5.5 Numerical Examples

In this section, we give examples making use of Padé-type approximants to 2π-periodic continuous-time signals $f(t)$.

Example 5.21. Let f be the signal $f(t) = |t|$ $(t \in \mathbb{R})$. The Fourier series of f into $[-\pi, \pi]$ is $\mathcal{F}(t) = (\pi/2) + \sum_{v=-\infty(v \neq 0)}^{\infty} [\{(-1)^v - 1\}/\pi v^2] e^{ivt}$. Define the $\mathbb{C}$-linear functional $T_f : \mathbb{P}(\mathbb{C}) \to \mathbb{C}$ associated with f by

$$T_f(x^v) := \begin{cases} \pi/2, & \text{if } v = 0 \\ \{(-1)^v - 1\}/\pi v^2, & \text{if } v \neq 0 \end{cases}$$

If $\mathcal{M} = (\pi_{m,k})_{m=0,1,\ldots;k=0,1,\ldots,m}$ is an infinite triangular interpolation matrix with complex entries $\pi_{m,k} \in D$, then a Padé-type approximation to the signal $f(t) = |t|$ is a function

$$(m/m+1)_f(t) = 2\mathrm{Re}\left(\left[e^{-it}T_f\left(\frac{V_{m+1}(e^{-it}) - V_{m+1}(x)}{e^{-it} - x}\right)\right] \Big/ V_{m+1}(e^{-it})\right) - (\pi/2)$$

where $V_{m+1}(x) = \prod_{k=0}^{m}(x - \pi_{m,k})$ is the generating polynomial of this approximation.

(i) If $m = 4$ and $\pi_{4,k}$ are the roots of the fifth Legendre polynomial in $[-1,1]$, then $V_5(x) = (63x^5 - 70x^3 + 15x)/8$ and therefore

$$(4/5)_f(t) = \frac{9094\pi\frac{5292}{\pi}\cos t - 10920\pi\cos 2t + \frac{12096}{\pi}\cos 3t + 1890\pi\cos 4t}{(63 - 70\cos 2t + 15\cos 4t)^2 + (-70\sin 2t + 15\sin 4t)^2} - \frac{\pi}{2}$$

$(t \in [-\pi,\pi])$. Indicatively, we have

t	$f(t)$	$(4/5)_f(t)$
$\pm\pi/2$	1.5707963	1.5707963
± 1	1	1.0957434
$\pm\pi/6$	0.5235987	0.5507175

(ii) If $m = 4$ and $\pi_{4,k} = 0$, then $V_5(x) = x^5$ and therefore

$$(4/5)_f(t) = \frac{\pi}{2} - \frac{4}{\pi}\cos t - \frac{4}{9\pi}\cos 3t$$

$(t \in [-\pi,\pi])$. Indicatively, we have

t	$f(t)$	$(4/5)_f(t)$
0	0	0
$\pm\pi/2$	1.5707963	1.5707963
$\pm\pi/8$	0.392699	0.3403377
$\pm\pi$	3.1415926	2.9855069
± 1	1	1.0756475
$\pm e$	2.7182818	2.673454

Example 5.22. Let f be the signal $f(t) = t^2$ $(t \in R)$. The Fourier series of f into $[-\pi,\pi]$ is $\mathcal{F}(t) = (\pi^2/3) + \sum_{\nu=-\infty(\nu=-\infty(\nu\neq 0)}^{\infty}[2(-1)^\nu/\nu^2]e^{i\nu t}$. Define the $\mathbb{C}$-linear functional $T_f : \mathbb{P}(\mathbb{C}) \to \mathbb{C}$ associated with f by

$$T_f(x^\nu) := \begin{cases} \pi^2/3, & \text{if } \nu = 0 \\ 2(-1)^\nu/\nu^2, & \text{if } \nu \neq 0. \end{cases}$$

If $\mathcal{M} = (\pi_{m,k})_{m=0,1,\ldots;k=0,1,\ldots,m}$ is an infinite triangular interpolation matrix with complex entries $\pi_{m,k} \in D$, then a Padé-type approximant to the signal $f(t) = t^2$ is a function

$$(m/m+1)_f(t) = 2\mathrm{Re}\left(\frac{e^{-it}T_f\left(\frac{V_{m+1}(e^{-it})-V_{m+1}(x)}{e^{-it}-x}\right)}{V_{m+1}(e^{-it})}\right) - \frac{\pi^2}{3}$$

where $V_{m+1}(x) = \prod_{k=0}^{m}(x - \pi_{m,k})$ is the generating polynomial of this approximation.

(i) If $m = 20$ and $\pi_{20,0} = \pi_{20,1} = \cdots = \pi_{20,19} = 0, \pi_{20,20} = -1$, then $V_{21}(x) = x^{21} + x^{20}$ and therefore

$$(20/21)_f(t) = 4\left(\sum_{v=0}^{19} \frac{(-1)^v}{v^2}\cos(vt) + \frac{1}{800}\frac{\cos(39t/2)}{\cos(t/2)}\right) + \frac{\pi^2}{3}$$

$(t \in [-\pi, \pi])$. Indicatively, we have

t	$f(t)$	$(4/5)_f(t)$
$\pm\pi/2$	2.4674011	2.466903
± 1	1	1.028211
$\pm\pi/6$	0.2741556	0.2649733

(ii) If $m = 40$ and $\pi_{40,k} = 0$, then $V_{41}(x) = x^{41}$ and therefore

$$(40/41)_f(t) = \frac{\pi^2}{3} - 4\sum_{v=0}^{40} \frac{(-1)^v}{v}\cos(vt)$$

$(t \in [-\pi, \pi])$. Indicatively, we have

t	$f(t)$	$(4/5)_f(t)$
0	0	0
$\pm\pi/2$	2.4674011	2.4653582
$\pm\pi/8$	0.1542125	0.1403446
$\pm\pi$	9.8696044	9.77084433
± 1	1	1.024185
$\pm e$	7.3890559	7.6610918

Example 5.23. Let f be the signal $f(t) = \frac{r\sin t}{1-2r\cos t+r^2}$ $(t \in \mathbb{R})$. The Fourier series of f into $[-\pi, \pi]$ is $\mathcal{F}(t) = \sum_{v=1}^{\infty}[ir^v/2]e^{-vt} + \sum_{v=1}^{\infty}[-ir^v/2]e^{ivt}$. Define the $\mathbb{C}$-linear functional $T_f : \mathbb{P}(\mathbb{C}) \to \mathbb{C}$ associated with f by

$$T_f(x^v) := \begin{cases} 0, & \text{if } v = 0 \\ -ir^v/2, & \text{if } v = 1, 2, \ldots. \end{cases}$$

If $\mathcal{M} = (\pi_{m,k})_{m=0,1,\ldots;k=0,1,\ldots,m}$ is an infinite triangular interpolation matrix with complex entries $\pi_{m,k} \in D$, then a Padé-type approximant to the signal $f(t)$ is a function

$$(m/m+1)_f(t) = 2\mathrm{Re}\left(e^{-it}T_f\left(\frac{V_{m+1}(e^{-it}) - V_{m+1}(x)}{e^{-it} - x}\right) \Big/ V_{m+1}(e^{-it})\right)$$

where $V_{m+1}(x) = \prod_{k=0}^{m}(x - \pi_{m,k})$ is the generating polynomial of this approximation.

(i) If $m = 14$ and $\pi_{14,14} = \pi_{14,14} = \cdots = \pi_{14,14} = 0$, then $(14/15)_f(t)$ is the trigonometric polynomial which equals the partial sum of the first fourteen terms in the Fourier expansion $\mathcal{F}(t)$ of $f(t)$ and indicatively, we have

t	$f_{1/8}(t)$	$(14/15)_t$	$f_{1/2}(t)$	$(14/15)_{f_{1/2}}(t)$	$f_{3/4}(t)$	$(14/15)_{f_{3/4}}(t)$
0	0	0	0	0	0	0
$\pi/3$	0.1215	0.1215	0.5773	0.5773	0.7994	0.8101
$-\pi$	0	0	0	0	0	0
$\pi/4$	0.1054	0.1053	0.6512	0.6512	1.056	1.055
$-\pi/6$	-0.0782	-0.0782	-0.6511	-1.1381	-1.4233	-1.4055
$\pi/5$	0.0903	0.0903	0.6664	0.9971	1.2632	1.2742
$\pi/2$	0.1231	0.1231	0.4	0.4000	0.48	0.4885

(ii) If $m = 3$ and $\pi_{3,k} = \cos[(2k+1)\pi/7]: \pi_{3,0} = 0.9009688$, $\pi_{3,1} = 0.2225209$, $\pi_{3,2} = -0.6234898$, $\pi_{3,3} = -1$, then $V_4(x) = x^4 + 0.5x^3 - x^2 - 0.375x + 0.125$ and therefore

$$(3/4)_f(t) = -r \times ([1.125r^2 - 0.3125r - 1.0625]\sin t$$

$$+ [-0.5r^2 - 1.125r - 1.3125]\sin 2t + [-r^2 + 0.5r + 1.375]\sin 3t)$$

$$\times ([1 + 0.5\cos t - \cos 2t - 0.375\cos 3t + 0.125\cos 4t]^2$$

$$+ [0.5\sin t - \sin 2t - 0.375\sin 3t + 0.125\sin 4t]^2)^{-1}$$

$(t \in [-\pi, \pi])$. In particular, there holds

t	$f(t)$	$(3/4)_f(t)$
0	0	0
$\frac{\pi}{3}$	$\frac{0.8660254r}{1-r-r^2}$	$\frac{0.8660254}{4.5468749}r(3.413294 + r)(1.113294 - r)$
$-\pi$	0	0
$\frac{\pi}{4}$	$\frac{0.7071067r}{1-1.4142135r-r^2}$	$\frac{r}{2.3765699}(r - 0.018477)(11.20945 - r)$
$-\frac{\pi}{6}$	$\frac{-0.5r}{1-1.7320508r+r^2}$	$-\frac{r}{1.5370791}(0.8705127r^2 + 0.6305285r + 0.2929083)$
$\frac{\pi}{5}$	$\frac{0.5877852r}{1-1.6180339r+r^2}$	$-\frac{r}{2.2188258}(0.7653264r^2 + 0.7775396r + 0.5654351)$
$\frac{\pi}{2}$	$\frac{r}{1+r^2}$	$-\frac{r}{5.28125}(r + 0.8967606)(1.2791136 - r)$

(iii) If $m = 3$ and $\pi_{3,0} = \pi_{3,1} = \pi_{3,2} = 0$ and $\pi_{3,3} = -i$, then $V_4(x) = x^4 + ix^3$ and therefore

$$(3/4)_f(t) = \frac{r}{2(1 - \sin t)} \times (-1 + 2\sin t + r\sin 2t + r^2 \sin 3t + r\sin 4t - r\cos t$$

$$+ [1 - r^2]\cos 2t + r\cos 3t)$$

($t \in [-\pi, \pi] \setminus \{\pi/2\}$). Observe that $(3/4)_f(t)$ is well defined everywhere on $-\pi, \pi]$, with the exception of the point $t = \pi/2$. This is consequence of our choice $\pi_{3,3} = -i$, which in particular implies $|\pi_{3,3}| = 1$ ($\Leftrightarrow \pi_{3,3} \in C$). However, it holds

t	$f(t)$	$(3/4)_f(t)$
0	0	$-0.5r^3$
$\frac{\pi}{3}$	$\frac{0.8660254r}{1-r-r^2}$	$3.7320507r(0.7320508 - 1.5r + 0.5r^2)$
$-\pi$	0	$0.5r(1 - r^2)$
$\frac{\pi}{4}$	$\frac{0.7071067r}{1-1.4142135r-r^2}$	$1.7071067r(0.4142135 - 0.4142135r + 0.7071067r^2)$
$-\frac{\pi}{6}$	$\frac{-0.5r}{1-1.7320508r+r^2}$	$0.3333333r(0.5 - 2.5980762r - 1.5r^2)$
$\frac{\pi}{5}$	$\frac{0.5877852r}{1-1.6180339r+r^2}$	$1.2129599r(0.4845874 + 2.0388417r - 0.3090169r^2)$
$\frac{\pi}{2}$	$\frac{r}{1+r^2}$	indefined

Example 5.24. Let f be the signal

$$f(t) = \begin{cases} 1, & \text{if } t = -\pi \\ 2(\frac{t}{\pi} + 1), & \text{if } -\pi < t \leq 0 \\ 2, & \text{if } 0 \leq t < 0 \\ 1, & \text{if } t = \pi \end{cases}$$

The Fourier series of f into $[-\pi, \pi]$ is $\mathcal{F}(t) = (3/2) + \sum_{v=-\infty(v\neq0)}^{\infty} \left[\frac{1-(-1)^v}{(v\pi)^2} + i\frac{(-1)^v}{v\pi} \right] e^{ivt}$. Define the $\mathbb{C}$-linear functional $T_f : \mathbb{P}(\mathbb{C}) \to \mathbb{C}$ associated with f by

$$T_f(x^v) := \begin{cases} 3/2, & \text{if } v = 0 \\ \frac{1-(-1)^v}{(v\pi)^2} + i\frac{(-1)^v}{v\pi}, & \text{if } v = 1, 2, \dots. \end{cases}$$

If $\mathcal{M} = (\pi_{m,k})_{m=0,1,\dots;k=0,1,\dots,m}$ is an interpolation matrix with complex entries $\pi_{m,k} \in D$, then a Padé-type approximant to the signal $f(t)$ is a function

$$(m/m+1)_f(t) = 2\text{Re}\left(\frac{e^{-it}T_f\left(\frac{V_{m+1}(e^{-it}) - V_{m+1}(x)}{e^{-it} - x} \right)}{V_{m+1}(e^{-it})} \right) - \frac{3}{2}$$

where $V_{m+1}(x) = \prod_{k=0}^{m}(x - \pi_{m,k})$ is the generating polynomial of the approximation.

(i) If $m = 3$ and $\pi_{3,0} = \pi_{3,1} = \pi_{3,2} = 0, \pi_{3,3} = -1/2$, then $V_4(x) = x^4 + (x^3/2)$ and therefore

$$(3/4)_f(t) = \frac{4}{5 + 4\cos t} \times (3.9526423 + 0.6366197\sin t$$
$$+ 3.5066059\cos t + 0.251163\cos 2t - 0.1061032\sin 3t$$
$$+ 0.0450316\cos 3t - 0.0506655\sin 4t) - 1.5$$

$(t \in [-\pi, \pi])$. In particular, we have

t	f	$(4/5)_f(t)$
$\pi/2$	2	2.0553617
$\pi/3$	2	2.0031648
$-\pi/3$	1.3333333	1.3730748
$\pi/5$	2	1.9466674
$-\pi/5$	1.6	1.5967140

(ii) If $m = 3$ and $\pi_{3,0} = \pi_{3,1} = \pi_{3,2} = \pi_{3,3} = 0$, then $V_4(x) = x^4$ and therefore

$$(3/4)_f(t) = 1.5 + 0.6366197\sin t + 0.4052847\cos t - 0.3183098\sin 2t$$
$$+ 0.2122065\sin 3t + 0.0450316\cos 3t$$

$(t \in [-\pi, \pi])$. In particular, we have

t	f	$(4/5)_f(t)$
$\pi/2$	2	1.9244132
$\pi/3$	2	1.9332752
$-\pi/3$	1.3333333	1.3819462
$\pi/5$	2	2.2890725
$-\pi/5$	1.6	1.5406813

References

1. Akhiezer, N. I.: The Classical Moment Problem. Hafner (1965)
2. Aoki, M.: State Space Modeling of Time Series. Springer-Verlag (1987)
3. Brezinski, C.: The asymptotic behavior of sequences and new series transformations based on the cauchy product. Rocky Mountain J. Math. **21**(1), 71–84 (1991)
4. Bromwich, T. J.: An Introduction to the Theory of Infinite Series, 2nd edn. Macmillan, London (1949)
5. Cazdow, J. A.: Spectral estimation: an overdetermined rational model equation approach. Proc. IEEE **70**, 907–939 (1982)
6. Carathéodory, C.: Über den Variabilitätsbereich der Koeffizienten von Potenzreihen, die gegebene Werte nicht annehmen. Math. Ann. **64**, 95–115 (1907)

7. Carathéodory, C.: Über den Variabilitätsbereich der Fourierschen Konstanten von positive harmonischen Functionen. Rend. di Palermo **32**, 193–217 (1911)

8. Clark, W. D.: Infinite series transformations and their applications. Thesis, University of Texas (1967)

9. Constantinescu, T.: Schur parameters, Factorization and Dilation Problems, Operator Theory: Advances and Applications, 2, Birkhäuser (1996)

10. Daras, N. J.: Rational approximation to harmonic functions. Numer. Algorithms **20**, 285–301 (1999)

11. Daras, N. J.: Padé and Padé-type approximation to 2π-periodic L^p functions. Acta Applicandae Mathematicae. **65**, 245–343 (2000)

12. Daras, N.J.: Interpolation methods for the evaluation of a 2π-periodic finite Baire measure. Appr. Theory Appl. **17**(2), 1–27 (2001)

13. Delsarte, Ph., Genin, Y., Kamp, Y., van Dooren, P.: Speech modeling and the trigonometric moment problem. Philips J. Res. **37**, 277–292 (1982)

14. Eiermann, M.: On the convergence of Padé-type approximatns to analytic functions. J. Comput. Appl. Math. **10**, 219–227 (1984)

15. Georgiou, T.T.: Realization of power spectra from partial covariance sequences. IEEE Trans. Acoustics, Speech and Signal Processing ASSP-35, 438–449 (1987)

16. Geronimo, J.: On polynomials orthogonal on the circle, on trigonometric moment problem, and on allied Carathéodory and Schur functions. C. R. (Doklady) Acad. Sci. USSR (N.S.) **39**, 291–295 (1943)

17. Geronimo, J.: On polynomials orthogonal on the circle, on trigonometric moment problem, and on allied Carathéodory and Schur functions. Mat. Sb. **105**(147) (1978), English transl. in Math. USSR Sb. **34**(2), 131–145 (1978)

18. Kailath, T.: Signal processing applications for some moment problems. In: Landau, H. (ed.) Moments in Mathematics, Amer. Math. Soc. **37**, 71–109 (1987)

19. Kalman, R.E.: Realization of Covariance Sequences. In: Proceedings of Toeplitz Memorial Conference (1981), Tel Aviv, Israel (1981)

20. Kay, S.M., Marple, S.L. Jr.: Spectrum analysis—A modern perspective. Proceedings IEEE **69**, 1380–1419 (1981)

21. Kimura, H.: Positive partial realization of covariance sequences. In: Byrnes, C.I., Lindquist, A. (eds.) Modeling, Identification and Robust Control, pp. 499–513. North-Holland (1987)

22. Lindquist, A., Picci, G.: On "subspace methods" Identification, in Systems and Networks. In: Hemke, U., Mennicken, R. Saurer, J. (eds.) Mathematical Theory and Applications II, Akademie Verlag, pp. 315–320 (1994)

23. Makhoul, J.: Linear prediction: a tutorial review. Proc. IEEE. **63**, 561–580 (1975)

24. Nevanlinna, R.: Über beschränkte Functionen, die in gegebenen Punken vorgeschiebene Werte annehmen. Ann. Acad. Sci. Fenn. Ser. A **13**(1), 1–71 (1919)

25. van Overschee, P., De Moor, B.: Subspace algorithms for stochastic identification problem. IEEE Trans. Automatic Contril AC-27, 382–387 (1982)

26. Schur, I.: On power series which are bounded in the interior of the unit circle. In: Gohberg, I. (ed.) Methods in Operator Theory and Signal Processing, Operator Theory: Advances and Applications, vol. 18, pp. 31–59 (1986) (Original (in German): J. Reine Angew. Math. **147**, 205–232 (1917))

27. Schur, I.: On power series which are bounded in the interior of the unit circle II. In: Gohberg, I. (ed.) Methods in Operator Theory and Signal Processing, Operator Theory: Advances and Applications, vol. 18, pp. 68–88 (1986) (Original (in German): J. Reine Angew. Math. **184**, 122–145 (1918))

28. Szegö, G.: Orthogonal polynomials. Amer. Math. Soc. Colloq. Publ. 23 (4th edition with revisions, 1975; 1st edition 1939), Amer. Math. Soc., Providence, RI

Part IV
Scientific Computing and Applications

Chapter 6
Computer Graphics Techniques in Military Applications

Dimitrios Christou, Antonios Danelakis, Marilena Mitrouli,
and Dimitrios Triantafyllou

Abstract The determination of intersection points of plane curves is a problem
of Computer Graphics with many applications in Applied Mathematics, Numerical
Analysis and many other scientific fields. More precisely, in military applications,
the trajectories of two flying objects such as missiles, aircrafts, etc. can be
interpreted by two plane curves. Our scope is to find the intersection points of
the given curves. The number of floating point operations (flops) of many classical
methods is not satisfactory, since they demand over $O(n^4)$ operations. Conversely,
many algorithms that are fast enough have serious problems with their numerical
stability. The main objective here is to develop fast and stable algorithms computing
the intersection points of plane curves. The error analysis and the computation of
complexity of all the proposed methods are analysed and demonstrated through
various examples.

D. Christou (✉)
School of Engineering and Mathematical Sciences, Systems and Control Centre, City University,
Northampton Square, London EC1V 0HB, UK
e-mail: Dimitrios.Christou.1@city.ac.uk

A. Danelakis
Department of Informatics and Telecommunications, University of Athens, Panepistimiopolis
15784, Athens, Greece
e-mail: a.danelakis@gmail.com

M. Mitrouli
Department of Mathematics, University of Athens, Panepistimiopolis 15784, Athens, Greece
e-mail: mmitroul@math.uoa.gr

D. Triantafyllou
Department of Military Sciences, Section of Mathematics and Engineering Sciences, Hellenic
Army Academy, 16675, Vari, Greece
e-mail: dtriant@math.uoa.gr

N.J. Daras (ed.), *Applications of Mathematics and Informatics in Military Science*,
Springer Optimization and Its Applications 71, DOI 10.1007/978-1-4614-4109-0_6,
© Springer Science+Business Media New York 2012

Keywords Intersection points • Plane curves • Symbolic-numeric computations
• Modified Sylvester matrix • Singular value decomposition

Mathematics Subject Classification (2010): 65D18, 68U05

6.1   Introduction

In many military applications the determination of intersection points of plane
curves is required. The main difficulties are the high complexity of the existing
algorithms and their numerical stability. In order to handle in an efficient way the
previous problem, Numerical Linear Algebra techniques will be used. Our aim
is to reduce the numerical complexity in a stable way. For achieving this aim,
we implement our method combining numerical and symbolical arithmetic in a
"hybrid" way.

Hybridity refers in its most basic sense to mixture: a mixture of different ways,
components, methods, etc. which can produce the same or similar results. The basic
idea of making something "hybrid" is to improve on its characteristics and therefore
make it work better. In our case, we focus on the mixture of symbolic arithmetic
and numeric arithmetic, which will be referred to as *hybrid arithmetic*. In a hybrid
arithmetic system both exact symbolic and numeric finite precision arithmetic
operations can be carried out simultaneously. *Symbolic computations* refer to
arithmetic operations either with arbitrary variables or fractions of integers to
represent the numerical input data. The symbolic computations which involve only
numerical data in rational format are also referred to as *rational computations* and
they are always performed in almost infinite accuracy, depending on the symbolic
kernel of the programming environment. Conversely, *numerical computations* refer
to arithmetic operations with numbers in floating-point format (decimal numbers).
However, the accuracy of the performed numerical computations is limited to
a specific number of decimal digits which gives rise to numerical rounding
errors that often cause serious complications and thus must be avoided. Recent
years have witnessed the emergence of new research combining symbolic and
numeric computations and leading to new kinds of algorithms, involving algebraic
computations with rational and approximate numeric arithmetic. This combination
gives a different perspective in the way to implement an algorithm and introduces
the notion of *hybrid computations*. Therefore, the different algebraic procedures,
which form an algorithm, can be implemented independently either using symbolic
computations or numerical computations. This kind of implementation is also
referred to as *hybrid implementation*. The hybridization of an algorithm (i.e.
the hybrid implementation of an algorithm) is possible in software programming
environments with symbolic-numeric arithmetic capabilities such as Maple, Matlab,
Mathematica and others which involve an efficient combination of symbolic (rational) and numerical (floating-point) operations. However, the effective combination

of symbolic and numerical operations depends on the nature of an algebraic method and the proper handling of the input data either as rational or floating-point numbers.

Considering the problem of intersection points, let C_t and C_u be two plane curves with parametric equations $P(t) = (x(t), y(t))$, $t \in [a,b]$ and $Q(u) = (\hat{x}(u), \hat{y}(u))$, $u \in [c,d]$, which represent the trajectories of a missile and a moving target. The computation of their intersection points requires three steps. In the first step, we transform one of these parametric equations of the curves to its implicit form. In the second step, we have to find the roots of a polynomial which form the possible intersection points of the curves. In the last step, we exactly specify the intersection points.

6.2 Symbolical Factorization of the Modified Sylvester Matrix

Theorem 6.1 ([5]). *Let* $P(t) = \left(\frac{u_1(t)}{v_1(t)}, \frac{u_2(t)}{v_2(t)} \right)$ *be a parametrization of a plane curve* C *with* $\gcd(u_1(t), v_1(t)) = \gcd(u_2(t), v_2(t)) = 1$. *Then the polynomial defining the implicit equation of* C *is*

$$F(x,y) = \mathrm{Res}_t \left(u_1(t) - x v_1(t), u_2(t) - y v_2(t) \right) = \mathrm{Res}_t \left(p(t), q(t) \right)$$

In order to compute the resultant of the polynomials $u_1(t) - x v_1(t)$ and $u_2(t) - y v_2(t)$, we will use the Sylvester matrix.

Definition 6.2 (The Sylvester Matrix [1]). Let a and b be two polynomials, respectively, of degree m and n, $n \leq m$. Thus:

$$a(s) = a_m s^m + a_{m-1} s^{m-1} + \cdots + a_1 s + a_0$$
$$b(s) = b_n s^n + b_{n-1} s^{n-1} + \cdots + b_1 s + b_0$$

The *Sylvester matrix* associated to a and b is an $(n+m) \times (n+m)$ matrix obtained as follows:

$$S = S(a,b) = \begin{bmatrix} a_m & a_{m-1} & a_{m-2} & \cdots & \cdots & a_0 & 0 & \cdots\cdots & 0 & 0 \\ 0 & a_m & a_{m-1} & a_{m-2} & \cdots & \cdots & a_0 & 0 & \cdots & 0 & 0 \\ \vdots & \vdots & \vdots & & & \vdots & \vdots & & & \vdots & \vdots \\ 0 & 0 & 0 & \cdots & a_m & a_{m-1} & \cdots & & \cdots & a_1 & a_0 \\ b_n & b_{n-1} & \cdots & & b_0 & 0 & 0 & \cdots & 0 & \cdots & 0 & 0 \\ 0 & b_n & \cdots & & & b_0 & 0 & \cdots & 0 & \cdots & 0 & 0 \\ \vdots & \vdots & \vdots & & & \vdots & \vdots & & & \vdots & \vdots \\ 0 & 0 & 0 & \cdots & \cdots & b_n & b_{n-1} & \cdots\cdots & b_1 & b_0 \end{bmatrix}$$

Collecting the two first rows of the blocks representing the polynomials $a(s)$ and $b(s)$, respectively, next collecting the second rows of them, etc., we construct a modified Sylvester matrix with n same $(2 \times (m+1))$ blocks.

$$S^*(a,b) = \begin{bmatrix} a_m & a_{m-1} & a_{m-2} & a_{m-3} & \cdots & a_{m-n} & a_{m-n-1} & a_{m-n-2} & \cdots & a_0 & 0 & 0 & \cdots & 0 \\ b_n & b_{n-1} & b_{n-2} & b_{n-3} & \cdots & b_0 & 0 & 0 & \cdots & 0 & 0 & 0 & \cdots & 0 \\ 0 & a_m & a_{m-1} & a_{m-2} & \cdots & a_{m-n+1} & a_{m-n} & a_{m-n-1} & \cdots & a_1 & a_0 & 0 & \cdots & 0 \\ 0 & b_n & b_{n-1} & b_{n-2} & \cdots & b_1 & b_0 & 0 & \cdots & 0 & 0 & 0 & \cdots & 0 \\ 0 & 0 & a_m & a_{m-1} & \cdots & a_{m-n+2} & a_{m-n+1} & a_{m-n} & \cdots & a_2 & a_1 & a_0 & \cdots & 0 \\ 0 & 0 & b_n & b_{n-1} & \cdots & b_2 & b_1 & b_0 & \cdots & 0 & 0 & 0 & \cdots & 0 \\ \cdot & \cdot & \cdot & \cdot & \cdots & \cdot & \cdot & \cdot & \cdots & \cdot & \cdot & \cdot & \cdots & \cdot \\ 0 & 0 & 0 & 0 & \cdots & 0 & a_m & a_{m-1} & \cdots & a_{m-n+1} & a_{m-n} & a_{m-n-1} & \cdots & a_0 \\ 0 & 0 & 0 & 0 & \cdots & 0 & b_n & b_{n-1} & \cdots & b_1 & b_0 & 0 & \cdots & 0 \\ 0 & 0 & 0 & 0 & \cdots & 0 & 0 & b_n & \cdots & b_2 & b_1 & b_0 & \cdots & 0 \\ \cdot & \cdot & \cdot & \cdot & \cdots & \cdot & \cdot & \cdot & \cdots & \cdot & \cdot & \cdot & \cdots & \cdot \\ \cdot & \cdot & \cdot & \cdot & \cdots & \cdot & \cdot & \cdot & \cdots & \cdot & \cdot & \cdot & \cdots & \cdot \\ 0 & 0 & 0 & 0 & \cdots & 0 & 0 & 0 & \cdots & 0 & 0 & b_n & \cdots & b_0 \end{bmatrix}$$

Let $a(t,x) = u_1(t) - x\upsilon_1(t)$ and $b(t,x) = u_2(t) - y\upsilon_2(t)$. For computing the resultant of these polynomials we have to triangularize their modified Sylvester matrix in respect of t. We zero the first element of the second row of $S^*(a,b)$ and we update the other entries using LU or QR factorization. Since the two first rows are repeated right shifted per one element, we update the rest of the matrix with no other calculations. Now we have $[\frac{n}{2}]$ same blocks, where $[\cdot]$ denotes the integer part of a real number. We continue similarly by zeroing and updating only one block at a time, until the whole matrix is triangularized. This procedure requires $O(n^2)$ flops, one order less than the complexity order of the classical LU or QR factorization. The numerical stability of the algorithm remains the same as that of the classical one. The determinant of S^* gives the implicit equation of the curve. Due to the term x in polynomials $a(t,x)$ and $b(t,x)$, the factorization of S^* must be implemented symbolically. The implicit equation of the curve will be

$$F(x,y) = \det(S^*(a,b))$$

where $\det(\cdot)$ denotes the determinant of a matrix.

6.3 Computing the Possible Intersection Points

If $F(x,y) = 0$ is the implicit equation of C_t, then by substituting the parametric equation of C_u into $F(x,y) = 0$, we obtain the equation:

$$F(\hat{x}(u), \hat{y}(u)) = 0$$

The possible intersection points of the curves are the roots of the above equation in $[c,d]$. Applying some steps of the bisection method and continuing with Newton-Raphson (N-R) or Secant method [2], we compute the previous equation. The convergence of N-R method is quadratic, but an initial value close to the root is required. Bisection converges slower than Newton–Raphson, but it is used in order to compute a "good" initial point for N-R.

6.4   Computing the Intersection Points

Let $p_1 = (x_1, y_1)$ be a possible intersection point of the two curves obtained in the previous step. If p_1 is an intersection point, then

$$p_1 = (x_1, y_1) = (x(t), y(t)) = P(t)$$

Computing t from the previous equation, if $t \in [a,b]$, then p_1 is an intersection point of the two curves, otherwise it is not. In order to implement this step, we will apply the Singular Value Decomposition (SVD) [3]. If (x_0, y_0) is a possible intersection point and M is the matrix obtained by substituting x_0 and y_0 in the modified Sylvester matrix, then the vector $(t^{m+n-1}, \ldots, t, 1)$ belongs to the nullspace of M. If the rank of M is $m+n-1$, then the vector $(t^{m+n-1}, \ldots, t, 1)$ is a basis of the nullspace of M. Since V is orthogonal, in the last column of that matrix in the SVD of M, we obtain a multiple of that vector with Euclidean norm $\alpha(t^{m+n-1}, \ldots, t, 1)$. Taking this into account the value of t corresponding to (x_0, y_0) is $t = \frac{\alpha t}{\alpha}$ [4]. In the first phase of SVD, we apply the modified QR factorization to the modified Sylvester matrix reducing the complexity.

6.5   Numerical Example

Specify the intersection points of the following curves $P(t), Q(u)$ [4] (Fig. 6.1):

Let $P(t) = (x(t), y(t))$ be a rational Bezier curve of degree 5, $t \in [0,1]$.

$$x(t) = \frac{v_1(t)}{w_1(t)} = \frac{-1/2(1-t)^5 - 5/4t^2(1-t)^3 + 5/4t^4(1-t) + 1/2t^5}{(1-t)^5 + 5/2t(1-t)^4 + 5/2t^2(1-t)^3 + 5/2t^3(1-t)^2 + 5/2t^4(1-t) + t^5}$$

$$y(t) = \frac{v_2(t)}{w_2(t)} = \frac{1/2(1-t)^5 - 5/4t(1-t)^4 - 5/4t^2(1-t)^3 + 5/4t^3(1-t)^2 + 5/4t^4(1-t) - 1/2t^5}{(1-t)^5 + 5/2t(1-t)^4 + 5/2t^2(1-t)^3 + 5/2t^3(1-t)^2 + 5/2t^4(1-t) + t^5}$$

and $Q(u) = (\bar{x}(u), \bar{y}(u))$ be a polynomial Bezier curve of degree 8, $u \in [0,1]$:

$$\bar{x}(u) = -24(1-t)^7 + 128t^2(1-t)^6 - 392t^3(1-t)^5 + 392t^5(1-t)^3 - 182t^6(1-t)^2 + 24t^7(1-t)$$

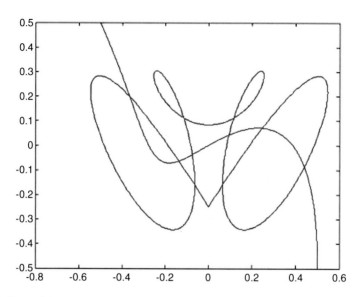

Fig. 6.1 Determining the intersection points of two parametric curves

$$\bar{y}(u) = -\tfrac{1}{4}(1-t)^8 + 32t(1-t)^7 - 476t^2(1-t)^6 + 1960t^3(1-t)^5 - 3010t^4$$
$$(1-t)^4 + 1960t^5(1-t)^3 - 476t^6(1-t)^2 + 32t^7(1-t) - \tfrac{1}{4}t^8$$

Then, we compute the polynomials $p = v_1(t) - xw_1(t)$ and $q = v_2(t) - yw_2(t)$:

$$p = -\tfrac{1}{2}(1-t)^5 - \tfrac{5}{4}t^2(1-t)^3 + \tfrac{5}{4}t^4(1-t) + \tfrac{1}{2}t^5 - $$
$$-x\left((1-t)^5 + \tfrac{5}{2}t(1-t)^4 + \tfrac{5}{2}t^2(1-t)^3 + \tfrac{5}{2}t^3(1-t)^2 + \tfrac{5}{2}t^4(1-t) + t^5\right)$$

$$q = \tfrac{1}{2}(1-t)^5 - \tfrac{5}{4}t(1-t)^4 - \tfrac{5}{4}t^2(1-t)^3 + \tfrac{5}{4}t^3(1-t)^2 + \tfrac{5}{4}t^4$$
$$(1-t) - \tfrac{1}{2}t^5 - y\left((1-t)^5 + \tfrac{5}{2}t(1-t)^4 + \tfrac{5}{2}t^2(1-t)^3 + \tfrac{5}{2}t^3(1-t)^2 + \tfrac{5}{2}t^4(1-t) + t^5\right)$$

The modified Sylvester matrix $S^*(p,q)$ of p and q in respect of t is

$$S = \begin{bmatrix}
1 & -5 & 35/4 & -(5x)/2-25/4 & (5x)/2+5/2 & -x-1/2 & 0 & 0 & 0 & 0 \\
-1 & 5/2 & -15/2 & 35/4-(5y)/2 & (5y)/2-15/4 & 1/2-y & 0 & 0 & 0 & 0 \\
0 & 1 & -5 & 35/4 & -(5x)/2-25/4 & (5x)/2+5/2 & -x-1/2 & 0 & 0 & 0 \\
0 & -1 & 5/2 & -15/2 & 35/4-(5y)/2 & (5y)/2-15/4 & 1/2-y & 0 & 0 & 0 \\
0 & 0 & 1 & -5 & 35/4 & -(5x)/2-25/4 & (5x)/2+5/2 & -x-1/2 & 0 & 0 \\
0 & 0 & -1 & 5/2 & -15/2 & 35/4-(5y)/2 & (5y)/2-15/4 & 1/2-y & 0 & 0 \\
0 & 0 & 0 & 1 & -5 & 35/4 & -(5x)/2-25/4 & (5x)/2+5/2 & -x-1/2 & 0 \\
0 & 0 & 0 & -1 & 5/2 & -15/2 & 35/4-(5y)/2 & (5y)/2-15/4 & 1/2-y & 0 \\
0 & 0 & 0 & 0 & 1 & -5 & 35/4 & -(5x)/2-25/4 & (5x)/2+5/2 & -x-1/2 \\
0 & 0 & 0 & 0 & -1 & 5/2 & -15/2 & 35/4-(5y)/2 & (5y)/2-15/4 & 1/2-y
\end{bmatrix}$$

We follow the next process:

- *Step 1: Symbolical computation of the determinant of $S^*(p,q)$*

We compute the determinant $F(x,y)$ of $S^*(p,q)$ by using the modified LU factorization for two polynomials. The special form of $S^*(p,q)$ and the use of symbolic operations guarantee a fast and error-free computation of the determinant which defines the implicit equation of $P(t)$.

$$F(x,y) = -\frac{507}{32}x^5 + \frac{22425}{128}x^4 - \frac{2145}{64}x^4 y - \frac{345}{32}y^2 x^3 - \frac{5625}{8}x^3 + \frac{975}{64}yx^3 + \frac{50625}{128}x^2 y - \frac{53025}{256}x^2 y^2 + \frac{58125}{1024}x^2 + \frac{165}{16}y^3 x^2 + \frac{20625}{256}y^2 x - \frac{136875}{1024}xy + \frac{33525}{256}y^3 x - \frac{15}{8}y^4 x + \frac{9375}{128}x - \frac{35625}{256}y^3 - \frac{69375}{512}y^2 - \frac{339}{64}y^5 - \frac{9375}{64}y - \frac{5325}{128}y^4$$

- *Step 2: Symbolic-numeric (hybrid) computation of the possible intersection points*

The polynomial, obtained after the symbolical substitution of $Q(u)$ into $F(x,y)$, has degree 40 with leading term $\frac{34373924217136700875 2927}{2048}u^{40}$ and its roots, which are computed numerically, are:

$$-0.1390025326, \ -0.01565456997, \ 0.006263135463, \ 0.02078384533,$$

$$0.2356938206, \ 0.7493929838, \ 0.9061596468, \ 0.9877287235$$

Six of the above roots are located in $[0,1]$ and hence there are six possible points in the intersection of the two curves:

$$x := -0.1370658337 \quad y := -0.06346706448$$
$$x := -0.3644703081 \quad y := 0.1968785911$$
$$x := -0.07523503040 \ y := -0.03785917025$$
$$x := 0.0931399242 \quad y := 0.04193972369$$
$$x := 0.4384598681 \quad y := -0.0852719548$$
$$x := 0.2453565161 \quad y := 0.0704612486$$

Substituting the first pair into S we take the matrix:

$$M = \begin{bmatrix}
1 & -5 & \frac{35}{4} & -5.907335416 & 2.157335416 & -0.3629341663 & 0 & 0 & 0 & 0 \\
0 & 1 & -5 & \frac{35}{4} & -5.907335416 & 2.157335416 & -0.3629341663 & 0 & 0 & 0 \\
0 & 0 & 1 & -5 & \frac{35}{4} & -5.907335416 & 2.157335416 & -0.3629341663 & 0 & 0 \\
0 & 0 & 0 & 1 & -5 & \frac{35}{4} & -5.907335416 & 2.157335416 & -0.3629341663 & 0 \\
0 & 0 & 0 & 0 & 1 & -5 & \frac{35}{4} & -5.907335416 & 2.157335416 & -0.3629341663 \\
-1 & 5/2 & -15/2 & 8.908667661 & -3.908667661 & 0.5634670645 & 0 & 0 & 0 & 0 \\
0 & -1 & 5/2 & -15/2 & 8.908667661 & -3.908667661 & 0.5634670645 & 0 & 0 & 0 \\
0 & 0 & -1 & 5/2 & -15/2 & 8.908667661 & -3.908667661 & 0.5634670645 & 0 & 0 \\
0 & 0 & 0 & -1 & 5/2 & -15/2 & 8.908667661 & -3.908667661 & 0.5634670645 & 0 \\
0 & 0 & 0 & 0 & -1 & 5/2 & -15/2 & 8.908667661 & -3.908667661 & 0.5634670645
\end{bmatrix}$$

• *Step 3: Numerical computation of the intersection points*

Now, computing the SVD of M by using the modified version of the QR method, the singular values and the last column of V are:

$$SV = \begin{bmatrix} 29.8054861571094812 \\ 20.7951855309906364 \\ 11.7147904374475332 \\ 5.87373827065510578 \\ 2.97903489165063552 \\ 1.27170139207374166 \\ 0.511452110138166161 \\ 0.419563134741653964 \\ 0.221314878081687760 \\ 1.29548396427539710 \times 10^{-12} \end{bmatrix} \text{ and } V = \begin{bmatrix} 0.00754085106243570884 \\ 0.0176876237818112404 \\ -0.0327212779420765210 \\ 0.0478347150969115134 \\ 0.0489506140418809604 \\ -0.0444343296751540593 \\ -0.0820511003903460212 \\ 0.176557015211754626 \\ 0.344819359122573276 \\ 0.913835716166440770 \end{bmatrix}$$

Then, $t_1 = \frac{V(9,10)}{V(10,10)} = 0.3773318913 \in [0,1]$ and so t_1 is a point in the intersection of the two curves. We continue similarly with the other five pairs (x,y).

With this process, we have actually achieved a significant reduction of the required flops and the corresponding computational time. Using the modified QR factorization in the first phase of the SVD for the bidiagonalization of M, the required flops are only $O(\frac{m^3}{2} + \frac{3}{2}mn^2 + 2m^2n - \frac{4}{3}n^3)$ (which for $m = n$ is equal to $\frac{8}{3}n^3$ flops) in contrast with the $O(\frac{4}{3}(m+n)^3)$ flops (which for $m = n$ is equal to $\frac{32}{3}n^3$ flops) of the classical QR bidiagonalization. The reduction of the flops in the first phase of the SVD is very important, since the total flops of SVD generally depend on the flops required for the bidiagonalization of the matrix.

6.6 Conclusions

In this paper, we have presented a method for computing the intersection points of plane curves. We modified classical methods, such as LU and QR factorization, in order to take advantage of the special form of the modified Sylvester matrix, reducing the numerical complexity of the methods. The modified QR factorization is implemented twice, one in the first step of the proposed algorithm and one in the third step (in the first phase of SVD). This computational technique reduced per one order the required complexity of the whole procedure, resulting in an algorithm of $O(n^3)$ flops. Our algorithm is numerically stable, since it uses slightly modified versions of QR factorization and SVD methods. The proposed algorithm was developed in a hybrid computational environment, where both numeric and the symbolic arithmetic were used in order to improve the quality of the given results. Finally, an analytical example regarding the computation of intersection points of plane curves was given.

Acknowledgement The fourth author (D.T.) acknowledges financial support from State Scholarships Foundation (IKY).

References

1. Barnett, S.: greatest common divisor from generalized sylvester resultant matrices. Linear and Multilinear Algebra. **8**, 271–279 (1980)
2. Conte, S.D., Carl de Boor: Elementary Numerical Analysis: An Algorithmic Approach, 3rd edn. pp. 74–80. McGraw-Hill Book Company, New York (1980)
3. Datta, B. N.: Numerical Linear Algebra and Applications, 2nd edn. SIAM, USA (2010)
4. Marco, A., Martinez, J-J.: A new source of structured singular value decomposition problems. Electronic Trans. Numer. Analy. **18**, 188–197 (2004)
5. Sendra, J.R., Winkler, F.: Tracing index of rational curve parameterizations. Comput. Aided Geomet. Des. **18**, 771–795 (2001)

Chapter 7
Numerical Optimization for the Length Problem

Christos Kravvaritis and Marilena Mitrouli

Abstract The length problem for normalized orthogonal (NO) matrices (satisfying $AA^T = A^T A = c(A)I_n$, for some constant c(A)), which is the determination of $c(n) = \sup\{c(A)|A \in \mathbb{R}^{n \times n}$, NO matrix$\}$, is formulated as a constrained optimization problem. The most appropriate numerical optimization technique for its study is analyzed. The corresponding numerical results provide useful experimental evidence concerning the possible values of $c(n)$ for various values of n and the relevant significance of Hadamard and weighing matrices is pointed out.

Keywords Length problem • Normalized orthogonal matrices • Numerical optimization

AMS Subject Classification: 65K05, 15B10, 15B34

7.1 Introduction

A *Hadamard matrix* H of order n is a matrix with entries ± 1 satisfying $HH^T = H^T H = nI_n$ [4,9,13]. It can be proved [7] that if H is a Hadamard matrix of order n then $n = 1,2$ or $n \equiv 0 \pmod 4$. However it is still an open conjecture whether Hadamard matrices exist for every n a multiple of 4. A more general class of orthogonal matrices are the *weighing matrices* W of order n and weight $n - k$ with entries $0, \pm 1$ satisfying $WW^T = W^T W = (n - k)I_n$ [4]. These matrices are special cases of a generalized class of orthogonal matrices called *orthogonal designs* [4].

C. Kravvaritis (✉) • M. Mitrouli
Department of Mathematics, University of Athens, Panepistemiopolis 15784, Athens, Greece
e-mail: ckrav@math.uoa.gr; mmitroul@math.uoa.gr

N.J. Daras (ed.), *Applications of Mathematics and Informatics in Military Science*,
Springer Optimization and Its Applications 71, DOI 10.1007/978-1-4614-4109-0_7,
© Springer Science+Business Media New York 2012

Hadamard and weighing matrices appear in several scientific fields, including the theory of Combinatorial Designs in Statistics, Coding Theory, Cryptography, Image and Signal Processing, and Analytical Chemistry [4, 9, 13]. Therefore they can be a powerful tool for military applications. For instance, in Coding Theory Hadamard matrices are used for constructing the error correcting Hadamard codes, which have good coding properties concerning the secure transmission of data sequences and provide high error correction rate. A famous application of the Hadamard code was in the NASA Mariner spacecraft missions in 1969 and 1972 to Mars, for correcting the picture transmission errors and for digitalizing and transmitting photos of Mars back to Earth. Since the messages from Mariner were fairly weak, the potential for errors was high, so high error correction capability was necessary. Also information transmission during recent flybys of the outer planets in the solar system is based on Hadamard matrices, too.

Hadamard matrices are used in Cryptography for guaranteeing the encryption, concealment, and secure transmission of data over a nonsecure channel; at personal level (e.g., safeguarding the PIN and banking transactions via internet) as well as at national level (e.g., national security, protection of medical and tax records and confidential data). Hadamard matrices are associated with bent functions which have the highest possible nonlinearity. Thus they effectively disguise and confuse characteristics of data sequences. This is a good property from a cryptographic point of view. Hadamard matrices and bent functions are used in the design of the so-called S-boxes, which are fundamental to the construction of cryptographically strong SPN algorithms (substitution-permutation-network) for private key cryptography.

7.1.1 The Length Problem

Hadamard and weighing matrices are examples of normalized orthogonal matrices. A matrix A is called *normalized* if $\max_{i,j} |a_{ij}| = 1$ and *normalized orthogonal* (NO) if A is normalized and satisfies additionally $AA^T = A^T A = c(A)I_n$ for some constant $c(A)$.

The problem of determining

$$c(n) = \sup\{c(A) | A \in \mathbb{R}^{n \times n}, \text{ NO matrix}\}$$

is called the *length problem* [3]. The term "length" is associated with the quantity $\sqrt{c(A)}$, which is the usual Euclidean length of every row of a NO matrix A. The NO matrices form a compact subset of the set of all $n \times n$ matrices $\mathbb{R}^{n^2}$ and the function $c(A)$ is continuous on it. Hence $c(n)$ exists and it is attained for a NO matrix A with $c(A) = c(n)$. Hadamard matrices are the only matrices known so far that attain length and growth [3, 11] equal to their order and also the maximum determinant value for matrices with entries ± 1, so it is intriguing to study their properties.

7.2 Optimization Formulation of the Length Problem

Here we present an approach that is based on the formulation of the length problem as an appropriate optimization problem. This strategy yields useful experimental results concerning the possible values of $c(n)$ for various n for NO matrices and helps to gather evidence relevant to the determination of $c(n)$.

The main idea of the method is to represent the n^2 unknown entries of an $n \times n$ matrix $X = (x_{ij})$, $1 \le i,j \le n$, as a vector $\bar{x}$ with elements $x_1, \ldots, x_{n^2}$. We apply the constraints

$$-1 \le x_i \le 1, \quad i = 1, \ldots, n^2$$

for the entries of a matrix X, since it is supposed to be normalized. Since maximizing f is equivalent to minimizing $-f$, one should minimize the objective function

$$-\sum_{j=1}^{n} x_{1j}^2,$$

or equivalently, in terms of the vector formulation,

$$f(\bar{x}) = -\sum_{i=1}^{n} x_i^2,$$

which is actually the negative of the square of the Euclidean length of the first row, corresponding to the quantity $c(A)$. This stand-alone property is not enough for the correct interpretation and modeling of the length problem, i.e., it cannot be formulated as an unconstrained optimization problem but constraints should be applied.

In order to write a proper optimization problem exploiting the specific formulation with this objective function, one should take also into account the equality of the usual Euclidean lengths of every two distinct rows, which are actually equal to $c(X)$. This property can be formulated in terms of x_{ij}, $1 \le i,j, \le n$, as

$$\sum_{j=1}^{n} x_{ij}^2 - \sum_{j=1}^{n} x_{i+1,j}^2 = 0, \quad 1 \le i \le n-1$$

and in the sequel in terms of $x_1, \ldots, x_{n^2}$ as

$$\sum_{k=1}^{n} x_{(i-1)n+k}^2 - \sum_{k=1}^{n} x_{in+k}^2 = 0, \quad 1 \le i \le n-1.$$

Furthermore the orthogonality of every two distinct rows of X must be taken into consideration. This fact leads to the constraints

$$\sum_{k=1}^{n} x_{ik} x_{jk} = 0, \quad 1 \le i < j \le n$$

and

$$\sum_{k=1}^{n} x_{(i-1)n+k} x_{(j-1)n+k} = 0, \quad 1 \leq i \leq n-1, \; i+1 \leq j \leq n,$$

formulated in terms of x_{ij} and x_i, respectively.

Summarizing, the optimization formulation of the length problem can be stated as follows:

$$\min_{\bar{x} \in \mathbb{R}^{n^2}} f(\bar{x}) = -\sum_{i=1}^{n} x_i^2 \tag{7.1}$$

$$\text{subject to } \sum_{k=1}^{n} x_{(i-1)n+k}^2 - \sum_{k=1}^{n} x_{in+k}^2 = 0, \quad 1 \leq i \leq n-1, \tag{7.2}$$

$$\sum_{k=1}^{n} x_{(i-1)n+k} x_{(j-1)n+k} = 0, \quad 1 \leq i \leq n-1, \; i+1 \leq j \leq n, \tag{7.3}$$

$$x_i + 1 \geq 0, \quad i = 1, \ldots, n^2 \tag{7.4}$$

$$\text{and } 1 - x_i \geq 0, \quad i = 1, \ldots, n^2. \tag{7.5}$$

So there are totally $n - 1 + \frac{n(n-1)}{2}$ equality constraints ($n - 1$ from (7.2) and $\frac{n(n-1)}{2}$ from (7.3)) and $2n^2$ inequality constraints. Since the relative computer program produces approximations to local constrained optima, depending on the choice of the initial point, and we are interested in global constrained optima, it is sensible to repeat the procedure for many initial points.

7.2.1 The Appropriate Optimization Technique

The most appropriate method for solving the optimization problem corresponding to the length problem is the sequential quadratic programming (SQP) algorithm. The SQP approach is one of the most effective methods for nonlinearly constrained optimization problems and it is equally suitable for both small and large problems. It is based on generation of steps by solving sequentially appropriately formulated quadratic subproblems. It can be adopted in both fundamental iterative strategies of optimization for moving from a current point x_k to a new iterate x_{k+1} in order to find a local minimum of an objective function: line search and trust-region frameworks. Here we illustrate the essential ingredients of a line search SQP algorithm for solving the following general nonlinear programming problem (7.6)–(7.8) with equality and inequality constraints

$$\min_{x \in \mathbb{R}^n} f(x) \tag{7.6}$$

$$\text{subject to } c_i(x) = 0, \quad i \in \mathcal{E}, \tag{7.7}$$

$$\text{and } c_i(x) \geq 0, \quad i \in \mathcal{I}, \tag{7.8}$$

where f, $c_i : \mathbb{R}^n \to \mathbb{R}$ are smooth. More details on SQP can be found in [1,5,10,12].

In a line search strategy, the algorithm chooses a direction p_k and searches along this direction from the current iterate x_k for a new iterate with a lower function value. The distance to move along p_k can be found by solving approximately the following one-dimensional minimization problem in order to find a step length a_k:

$$\min_{a>0} f(x_k + ap_k).$$

Then the new iterate is updated as $x_{k+1} = x_k + a_k p_k$. The procedure is terminated when a suitable criterion is satisfied, e.g., $\|\nabla f(x_k)\|$ is sufficiently small.

The essential idea of SQP is to model the problem (7.6)–(7.8) at the current iterate x_k by a quadratic programming subproblem and to use the minimizer of this subproblem for defining a new iterate x_{k+1}. It is important to design the quadratic subproblem so that it yields a good step for the underlying constrained optimization problem and so that the overall SQP algorithm has good convergence properties and good practical performance. The quadratic programming subproblem is obtained by linearizing both the inequality and equality constraints:

$$\min_{p\in\mathbb{R}^n} \frac{1}{2} p^T B_k p + \nabla f_k^T p$$

$$\text{subject to } \nabla c_i(x_k)^T p + c_i(x_k) = 0, \quad i\in\mathcal{E},$$

$$\text{and } \nabla c_i(x_k)^T p + c_i(x_k) \geq 0, \quad i\in\mathcal{I}.$$

B_k stands for $B(x_k, \lambda_k)$, where $B(x,\lambda) = \nabla^2_{xx}\mathcal{L}(x,\lambda)$ is the Hessian matrix of the Lagrangian. This quadratic problem can be solved by means of any algorithm for quadratic programming described in [1,12], with most preferable being an active set strategy similar to that of [5]. The solution of the quadratic subproblem produces a vector p_k, which is used to form a new iterate $x_{k+1} = x_k + a_k p_k$.

7.3 Numerical Results

After using the appropriate computer package repeatedly and for many initial points we obtained the following numerical results, which confirm the theoretically known values $c(3) = 9/4$, $c(4) = 4$ and the experimental results carried out in [3] for $n = 5, 6, 7$ and also extend them for the cases $n = 8, \ldots, 16$. Precisely, we are led to believe that the results given in Table 7.1 hold, where the values of $c(n)$ for n odd are rounded to four decimal digits. First of all we mention that the algorithm exhibits a very stable behavior, i.e., for the majority of the many and various random initial points chosen the results of Table 7.1 occur.

These results give rise to the following observations. The values obtained for $c(n)$, n even, are given for the special cases of orthogonal designs presented before. More specifically, for $n \equiv 0 \bmod 4$, $c(n)$ is always given by a Hadamard matrix of

Table 7.1 Numerical results
for the length problem

n	$c(n)$
5	3.3611
6	5
7	5.0777
8	8
9	6.4308
10	9
11	8.4495
12	12
13	10.3934
14	13
15	11.8511
16	16

order n and equals n. For $n \equiv 2 \bmod 4$, $c(n)$ is always given by a weighing matrix of order n and weight $n - 1$ and equals $n - 1$. These observations can be proved also theoretically yielding the following Propositions 7.1 and 7.2.

Proposition 7.1. $c(n) = n$ *iff there exists a Hadamard matrix of order n.*

Proof. We deal with NO matrices, so it is clear that $c(n) \leq n$. Because of the NO property, the only possibility to obtain the maximum value $c(n) = n$ is when we have a matrix with mutually orthogonal distinct rows and columns, and the inner product of every row and column with itself is equal to n. The only possibility are the Hadamard matrices.   □

The second case can be proved by means of the same argumentation.

Proposition 7.2. $c(n) = n - 1$ *iff there exists a weighing matrix of order n and weight $n - 1$.*

The rest of the values do not seem to follow a specific, predictable pattern and it is interesting to examine whether the values in Table 7.1 for n odd are true or not, and whether they are subject to any particular formula, which would help to predict $c(n)$ for larger values of n. The above results provide also evidence that $c(n)$ seems to be increasing for n odd and even *separately*, whereas the monotonicity of $c(n)$ for arbitrary values of n does not become apparent. Furthermore it is questionable whether $c(5) < c(4)$, $c(11) < c(10)$, $c(13) < c(12)$, etc.

7.4 Conclusions

The length problem has been defined and the role of Hadamard and weighing matrices has been highlighted. Its interpretation as an appropriately formutated constrained nonlinear optimization problem yields interesting experimental results offering insight into the possible values of $c(n)$. It is always useful to formulate some problems as appropriately defined optimization problems, if possible, as it was done, e.g., in [6]. The corresponding numerical results led to proving that the inequality part in the famous Cryer's complete pivoting conjecture for the growth factor in Gaussian Elimination [2, 3, 8], which was believed for many years to be true, is finally false. Ongoing research is focused on whether the values, which don't correspond to Hadamard and weighing matrices, follow a specific, predictable pattern.

References

1. Bazaraa, M.S., Shetty, C.M.: Nonlinear Programming, theory and algorithms. Wiley, Toronto (1979)
2. Cryer, C.W.: Pivot size in gaussian elimination. Numer. Math. **12**, 335–345 (1968)
3. Day, J., Peterson, B.: Growth in gaussian elimination. Amer. Math. Monthly **95**, 489–513 (1988)
4. Geramita, A.V., Seberry, J.: Orthogonal Designs: Quadratic Forms and Hadamard Matrices. Marcel Dekker, New York-Basel (1979)
5. Gill, P.E., Murray, W., Wright, M.H.: Practical Optimization. Academic Press, London (1981)
6. Gould, N.: On growth in Gaussian elimination with pivoting. SIAM J. Matrix Anal. Appl. **12**, 354–361 (1991)
7. Hadamard, J.: Résolution d'une question relative aux déterminants. Bull. Sci. Math. **17**, 240–246 (1893)
8. Higham, N.J.: Accuracy and Stability of Numerical Algorithms. SIAM, Philadelphia (2002)
9. Horadam, K.J.: Hadamard matrices and their applications. Princeton University Press, Princeton (2007)
10. Jarre, F., Stoer, J.: Optimierung. Springer, Berlin (2004)
11. Kravvaritis, C., Mitrouli, M.: The growth factor of a Hadamard matrix of order 16 is 16. Numer. Linear Algebra Appl. **16**, 715–743 (2009)
12. Nocedal, J., Wright, S.J.: Numerical Optimization. Springer, New York (1999)
13. Wallis, W.D., Street A.P., Wallis, J.S.: Combinatorics: Room Squares, Sum-Free Sets, Hadamard Matrices, Lectures notes in mathematics, vol. 292. Springer, New York (1972)

Part V
Simulation and Combat Models

Chapter 8
Adaptive Policies for Sequential Sampling under Incomplete Information and a Cost Constraint

Apostolos Burnetas and Odysseas Kanavetas

Abstract We consider the problem of sequential sampling from a finite number of independent statistical populations to maximize the expected infinite horizon average outcome per period, under a constraint that the expected average sampling cost does not exceed an upper bound. The outcome distributions are not known. We construct a class of consistent adaptive policies, under which the average outcome converges with probability 1 to the true value under complete information for all distributions with finite means. We also compare the rate of convergence for various policies in this class using simulation.

Keywords Stochastic learning and adaptive control • Sequential design • Sampling cost constraint

AMS Subject Classification: Primary 93E35, Stochastic learning and adaptive control; Secondary 62L05, Sequential designs

8.1   Introduction

In this paper we consider the problem of sequential sampling from k independent statistical populations with unknown distributions. The objective is to maximize the expected outcome per period achieved over infinite horizon, under a constraint that the expected sampling cost per period does not exceed an upper bound. The introduction of a sampling cost introduces a new dimension in the standard trade-off between experimentation and profit maximization faced in problems of control under incomplete information. The sampling cost may prohibit using

A. Burnetas (✉) • O. Kanavetas
Department of Mathematics, University of Athens,
Panepistemiopolis 15784, Athens, Greece
e-mail: aburnetas@math.uoa.gr; okanav@math.uoa.gr

N.J. Daras (ed.), *Applications of Mathematics and Informatics in Military Science*,
Springer Optimization and Its Applications 71, DOI 10.1007/978-1-4614-4109-0_8,
© Springer Science+Business Media New York 2012

populations with high mean outcomes because their sampling cost may be too high. Instead, the decision maker must identify the subset of populations with the best combination of outcome versus cost and allocate the sampling effort among them in an optimal manner.

From the mathematical point of view, this class of problems incorporates statistical methodologies into mathematical programming problems. Indeed, under complete information, the problem of effort allocation under cost constraints is typically formulated in terms of linear or nonlinear programming. However when some of the problem parameters are not known in advance but must be estimated by experimentation, the decision maker must design adaptive learning and control policies that ensure learning about the parameters while at the same time ensuring that the profit sacrificed for the learning process is as low as possible.

The model in this paper falls in the general area of multi-armed bandit problems, which was initiated by [10], who proposed a simple adaptive policy for sequentially sampling from two unknown populations in order to maximize the expected outcome per unit time infinite horizon. Lai and Robbins [6] generalize the results by constructing asymptotically efficient adaptive policies with optimal convergence rate of the average outcome to the optimal value under complete information and show that the finite horizon loss due to incomplete information increases with logarithmic rate. Katehakis and Robbins [4] prove that simpler index-based efficient policies exist in the case of normal distributions with unknown means, while Burnetas and Katehakis [2] extend the results on efficient policies in the nonparametric case of discrete distributions with known support.

In a finite horizon Kulkarni and Lugosi [5] develop a minimax version of the [6] results for two populations, while Auer et al. [1] construct policies which also achieve logarithmic regret uniformly over time, rather than only asymptotically.

In all works mentioned above there is no side constraint in sampling. Problems with adaptive sampling and side constraints are scarce in the literature. Wang [11] considers a multi-armed bandit model with constraints and adopts a Bayesian formulation and the Gittins-index approach. The paper proposes several heuristic policies. Pezeshk and Gittins [8] also consider the problem of estimating the distribution of a single population with sampling cost under the assumption that the number of users who will benefit from the depends on the outcome of the estimation. Finally, Madani et al. [7] present computational complexity analysis for a version of the multi-armed bandit problem with Bernoulli outcomes and Beta priors, where there is a budget available for sampling.

Another approach, which is closer to the one we adopt here, is to consider the family of stochastic approximations and reinforcement learning algorithms. The general idea is to select the sampled population following a randomized policy with randomization probabilities that are adaptively modified after observing the outcome in each period. The adaptive scheme is based on the stochastic approximation algorithm. Algorithms of this type are analyzed in [9] for the more general case where the population outcomes have Markovian dynamics instead of being i.i.d.

The contribution of this paper is the construction of a family of policies for which the average outcome per period converges to the optimal value under complete information for all distributions of individual populations with finite means. In this sense, it generalizes the results of [10] by including a sampling cost constraint. The paper is organized as follows. In Sect. 8.2, we describe the model in the complete and incomplete information framework. In Sect. 8.3, we construct a class of adaptive sampling policies and prove that it is consistent. In Sect. 8.4, we explore the rate of convergence of the proposed policies using simulation. Section 8.5 concludes.

8.2 Model Description

Consider the following problem in adaptive sampling. There are k independent statistical populations, $i = 1, \ldots, k$. Successive samples from population i constitute a sequence of i.i.d. random variables $X_{i1}, X_{i2}, \ldots$ following a univariate distribution with density $f_i(\cdot)$ with respect to a nondegenerate measure v. Then the stochastic model is uniquely determined by the vector $f = (f_1, \ldots, f_k)$ of individual pdfs. Given f let $\mu(f)$ be the vector of expected values, i.e., $\mu_i(f) = E^{f_i}(X_i)$. The form of f is not known. In each period the experimenter must select a population to obtain a single sample from. Sampling from population i incurs cost c_i per sample and without loss of generality we assume $c_1 \leq c_2 \leq \cdots \leq c_k$, but not all equal. The objective is to maximize the expected average reward per period subject to the constraint that the expected average sampling cost per period over infinite horizon does not exceed a given upper bound C_0. Without loss of generality we assume $c_1 \leq C_0 < c_k$. Indeed if $C_0 < c_1$ then the problem is infeasible. On the other hand if $C_0 \geq c_k$ then the cost constraint is redundant. Let $d = max\{j : c_j \leq C_0\}$. Then $1 \leq d < k$ and $c_d \leq C_0 < c_{d+1}$.

8.2.1 Complete Information Framework

We first analyze the complete information problem. If all $f_i(\cdot)$ are known, then the problem can be modeled via linear programming. Consider a randomized sampling policy which at each period selects population j with probability x_j, for $j = 1, \ldots, k$. To find a policy that maximizes the expected reward, we can formulate the following linear program in standard form:

$$z^* = \max \sum_{j=1}^{k} \mu_j x_j$$

$$\sum_{j=1}^{k} c_j x_j + y = C_0$$

$$\sum_{j=1}^{k} x_j = 1$$

$$x_j \geq 0, \forall j. \tag{8.1}$$

Note that z^* depends on f only through the vector $\underline{\mu}(f)$, i.e., z^* is the same for all collections of pdf with the same μ. Therefore in the remainder we will denote z^* as a function of the unknown mean vector $\underline{\mu}$.

In the analysis we will also use the dual linear program (DLP) of (8.1),

$$z_D^* = \min \ g + C_0 \lambda$$

$$g + c_1 \lambda \geq \mu_1$$

$$\vdots$$

$$g + c_k \lambda \geq \mu_k$$

$$g \in \mathbb{R}, \lambda \geq 0,$$

with two variables λ and g which correspond to the first and second constraints of (8.1), respectively.

The basic matrix B corresponding to a basic feasible solution (BFS) of problem (8.1) may take one of two forms:

In the first case, the basic variables are x_i, x_j, for two populations i, j, with $c_i \leq C_0 \leq c_j, c_i < c_j$, and the basic matrix is

$$B = \begin{pmatrix} c_i & c_j \\ 1 & 1 \end{pmatrix}.$$

The BFS is then

$$x_i = \frac{c_j - C_0}{c_i - c_j}, \ x_j = \frac{C_0 - c_i}{c_i - c_j}, \ \text{and} \ x_m = 0 \ \text{for} \ m \neq i, j, \ y = 0,$$

with

$$z(\underline{x}) = \mu_i x_i + \mu_j x_j.$$

The solution is nondegenerate when $c_i < C_0 < c_j$ and degenerate when $C_0 = c_i$ or $C_0 = c_j$. In the latter case, it corresponds to sampling from a single population $l = i$ or $l = j$, respectively:

$$x_l = 1, \ x_m = 0 \ \forall m \neq l, \ y = 0,$$

with

$$z(\underline{x}) = \mu_l.$$

The second case of a BFS corresponds to basic variables x_i, y for a population i with $c_i \leq C_0$. The basic matrix is

$$\mathbf{B} = \begin{pmatrix} c_i & 1 \\ 1 & 0 \end{pmatrix}.$$

In this case the BFS corresponds to sampling from population i only

$$x_i = 1, \ x_m = 0 \ \forall m \neq i, \ y = C_0 - c_i,$$

with

$$z(\underline{x}) = \mu_i.$$

The solution is nondegenerate if $c_i < C_0$; otherwise it is degenerate.

From the above it follows that a BFS is degenerate if $x_l = 1$ for some l with $c_l = C_0$. Any basic matrix B that includes x_l as a basic variable corresponds to this BFS.

For a BFS x let

$$b = \{i : x_i > 0\}.$$

Then, either $b = \{i, j\}$ for some i, j with $i \leq d \leq j$, or $b = \{i\}$ for some $i \leq d$. There is a one to one correspondence between BFSs and sets b of this form. We use K to denote the set of BFS, or equivalently

$$K = \{b \ : \ b = \{i, j\}, \ i \leq d \leq j \text{ or } b = \{i\}, \ i \leq d\}.$$

Since the feasible region of (8.1) is bounded, K is finite.

For a basic matrix B, let $v^B = (\lambda^B, g^B)$ denote the dual vector corresponding to B, i.e., $v^B = \mu_B B^{-1}$, where $\mu_B = (\mu_i, \mu_j)$, or $\mu_B = (\mu_i, 0)$, depending on the form of B.

Regarding optimality, a BFS is optimal if and only if for at least one corresponding basic matrix B the reduced costs (dual slacks) are all nonnegative:

$$\phi_a^B \equiv c_a \lambda^B + g^B - \mu_a \geq 0, \ \alpha = 1, \ldots, k.$$

A basic matrix B satisfying this condition is optimal. Note that if an optimal BFS is degenerate, then not all basic matrices corresponding to it are necessarily optimal.

It is easy to show that the reduced costs can be expressed as a linear combination $\phi_a^B = \underline{w}_a^B \underline{\mu}$, where $\underline{w}_a^B$ is an appropriately defined vector that does not depend on $\underline{\mu}$.

We finally define the set with optimal solutions of (8.1) for a $\underline{\mu}$,

$$s(\underline{\mu}) = \{b \in K : b \text{ corresponds to an optimal BFS}\}.$$

An optimal solution of (8.1) specifies randomization probabilities that guarantee maximization of the average reward subject to the cost constraint. Note that an alternative way to implement the optimal solution, without randomization, is to sample periodically from all populations so that the proportion of samples from each population j is equal to x_j. This characterization of a policy is valid if randomization probabilities are rational.

8.2.2   Incomplete Information Framework

In this paper we assume that the population distributions are unknown. Specifically we make the following assumption.

Assumption 1 *The outcome distributions are independent, and the expected values* $\mu_\alpha = E(X_\alpha) < \infty$, $\alpha = 1, \ldots, k$.

Let F be the set of all $f = (f_1, \ldots, f_k)$ which satisfy Assumption 1 Class F is the effective parameter set in the incomplete information framework. Under incomplete information, a policy as that in Sect. 8.2.1, which depends on the actual value of $\underline{\mu}$, is not admissible. Instead we restrict our attention to the class of adaptive policies, which depend only on the past observations of selections and outcomes.

Specifically, let A_t, X_t, $t = 1, 2, \ldots$ denote the population selected and the observed outcome at period t. Let $h_t = (\alpha_1, x_1, \ldots, \alpha_{t-1}, x_{t-1})$ be the history of actions and observations available at period t.

An adaptive policy is defined as a sequence $\pi = (\pi_1, \pi_2, \ldots)$ of history dependent probability distributions on $\{1, \ldots, k\}$, such that

$$\pi_t(j, h_t) = P(A_t = j | h_t).$$

Given the history h_n, let $T_n(\alpha)$ denote the number of times population α has been sampled during the first n periods

$$T_n(\alpha) = \sum_{t=1}^{n} 1\{A_t = \alpha\}.$$

Let S_n^π be the reward up to period n:

$$S_n^\pi = \sum_{t=1}^{n} X_t,$$

and C_n^π be the total cost up to period n:

$$C_n^\pi = \sum_{t=1}^{n} c_{A_t}.$$

These quantities can be used to define the desirable properties of an adaptive policy, namely feasibility and consistency.

Definition 8.1.   A policy π is called feasible if

$$\limsup_{n \to \infty} \frac{E^\pi(C_n^\pi)}{n} \le C_0, \ \forall f \in F. \tag{8.2}$$

Definition 8.2. A policy π is called consistent if it is feasible and

$$\lim_{n \to \infty} \frac{S_n^\pi}{n} = z^*(\underline{\mu}), \text{ a.s. } \forall f \in F.$$

Let Π^F and Π^C denote the class of feasible and consistent policies, respectively. The above properties are reasonable requirements for an adaptive policy. The first ensures that the long-run average sampling cost does not exceed the budget. The second definition means that the long-run average outcome per period achieved by π converges with probability one to the optimal expected value that could be achieved under full information, for all possible population distributions satisfying Assumption 1.

Note that consistency as defined in Definition 8.2 is equivalent to the notion of strong consistency of an estimator function.

8.3 Construction of a Consistent Policy

A key question in the incomplete information framework is whether feasible and, more importantly, consistent policies exist and how they can be constructed.

It is very easy to show that feasible policies exist, since the sampling costs are known. Indeed any randomized policy, such as those defined in Sect. 8.2.1, with randomization probabilities satisfying the constraints of LP (8.1) is feasible for any distribution f. Thus, $\Pi^F \neq \emptyset$.

On the other hand, the construction of consistent policies is not trivial. A consistent policy must accomplish three goals: first to be feasible, second to be able to estimate the mean outcomes from all populations, and third, in the long run, to sample from the nonoptimal populations rarely enough so as not to affect the average profit.

In this section we establish the existence of a class of consistent policies. The construction follows the main idea of [10], based on sparse sequences, which is adapted to ensure feasibility.

We start with some definitions. For any population j, let $\hat{\mu}_{j,t}$, $t = 1, 2, \ldots$ be a strongly consistent estimator of μ_j, i.e., $\lim_{t \to \infty} \hat{\mu}_{j,t} = \mu_j$ a.s. $-f_j$. Such estimators exist; for example from Assumption 1, the sample mean $\overline{X}_{j,t} = \frac{1}{t} \sum_{k=1}^{t} X_{j,k}$ is strongly consistent.

For any n, let $\underline{\hat{\mu}}_n = (\hat{\mu}_{j,T_j(n)}, j = 1, \ldots, k)$ be the vector estimates of $\underline{\mu}$ based on the history up to period n. Also let $\hat{z}_n = z(\underline{\hat{\mu}}_n)$ denote the optimal value of the linear program in (8.1) where the estimates are used in place of the unknown mean vector in the objective. $\hat{z}_n$ will be referred to as the Certainty-Equivalence LP. Note that $s(\underline{\hat{\mu}}_n)$ is the set of optimal BFS of $\hat{z}_n$.

The solution of $\hat{z}_n$ corresponds to a sampling policy determined by an optimal vector $\hat{x}_n$, so that $\hat{z}_n = \underline{\hat{\mu}}_n' \hat{x}_n$.

We next define a class of sampling policies, which we will show to be consistent. Consider k nonoverlapping sparse sequences of positive integers,

$$\tau_j = \{\tau_{j,m}, \ m = 1, 2, \ldots\}, \ j = 1, \ldots, k,$$

such that

$$\lim_{m \to \infty} \frac{\tau_{j,m}}{m} = \infty, \ j = 1, \ldots, k. \tag{8.3}$$

Now define policy π^0 which in period n selects any population j with probability equal to

$$\pi^0(j|h_n) = \begin{cases} 1, & \text{if } \tau_{j,m} = n \text{ for some } m \geq 1 \\ \hat{x}_{n,j}, & \text{otherwise} \end{cases}$$

where $\hat{x}_n$ is any optimal BFS of the certainty-equivalence LP $\hat{z}_n$.

The main idea in π^0 is that at periods which coincide with the terms of sequence τ_j, population j is selected regardless of the history. These instances are referred to as forced selections of population j. The purpose of forced selections is to ensure that all populations are sampled infinitely often, so that the estimate vector $\hat{\underline{\mu}}_n$ converges to the true mean $\underline{\mu}$ as $n \to \infty$.

On the other hand, because sequences τ_j are sparse, the fraction of forced selections periods converges to zero for all j, so that sampling from the nonoptimal populations does not affect the average outcome in the long run.

In the remaining time periods, which do not coincide with a sparse sequence term, the sampling policy is that suggested by the certainty equivalence LP, i.e., the experimenter in general randomizes between those populations, which, based on the observed history, appear to be optimal.

In the next theorem we prove the main result of the paper, namely that of $\pi^0 \in \Pi^C$. The proof adapts the main idea of [10] to the problem with the cost constraint.

Theorem 8.3. *Policy π^0 is consistent.*

Before we show Theorem 8.3, we prove an intermediate result which shows that if in some period the certainty equivalence LP yields an optimal solution that is nonoptimal under the true distribution f, then the estimate of at least one population mean must be sufficiently different from the true value. We use the supremum norm $\|x\| = \max_j |x|$.

Lemma 8.4. *For any $\underline{\mu}$ there exists $\varepsilon > 0$ such that for any $n = 1, 2, \ldots$ if $b \in s(\hat{\underline{\mu}}_n)$ and $b \notin s(\underline{\mu})$ for some $b \in K$, then $\|\underline{\mu} - \hat{\underline{\mu}}_n\| \geq \varepsilon$.*

Proof. Since $b \notin s(\underline{\mu})$, we have that for any basic matrix B' corresponding to BFS b there exists at least one $m \in \{1, \ldots, k\}$ such that $\phi_m^{B'}(\underline{\mu}) < 0$. Therefore,

$$-\underline{w}_m^{B'} \underline{\mu} = -\phi_m^{B'}(\underline{\mu}) > 0. \tag{8.4}$$

In addition, since $b \in s(\underline{\hat{\mu}}_{-n})$, there exists a basic matrix B corresponding to b, such for any $m \in \{1,\ldots,k\}$ it is true that $\phi_m^B(\underline{\hat{\mu}}_{-n}) \geq 0$, thus,

$$\underline{w}_m^B \underline{\hat{\mu}}_{-n} = \phi_m^B(\underline{\hat{\mu}}_{-n}) \geq 0. \tag{8.5}$$

For this basic matrix B, it follows from (8.4) and (8.5) that

$$\underline{w}_m^B \underline{\hat{\mu}}_{-n} - \underline{w}_m^B \underline{\mu} \geq -\phi_m^B(\underline{\mu}) = |\phi_m^B(\underline{\mu})| > 0$$

$$\Rightarrow \underline{w}_m^B(\underline{\hat{\mu}}_{-n} - \underline{\mu}) \geq |\phi_m^B(\underline{\mu})|$$

$$\Rightarrow k\|\underline{w}_m^B\| \|\underline{\hat{\mu}}_{-n} - \underline{\mu}\| \geq |\phi_m^B(\underline{\mu})|$$

$$\Rightarrow \|\underline{\hat{\mu}}_{-n} - \underline{\mu}\| \geq \frac{|\phi_m^B(\underline{\mu})|}{k\|\underline{w}_m^B\|},$$

because from the property $\underline{w}_m^B \underline{\mu} < 0$ it follows that $\|\underline{w}_m^B\| > 0$.

Now let

$$\varepsilon = \min_{b \in K, b \notin s(\underline{\mu})} \min_{B \in b} \min_{m \in \{1,\ldots,k\}} \left\{ \frac{|\phi_m^B(\underline{\mu})|}{k\|\underline{w}_m^B\|} : \phi_m^B(\underline{\mu}) < 0 \right\} > 0.$$

where the minimization over $B \in b$ is taken over all basic matrices corresponding to BFS b.

Then $\|\underline{\hat{\mu}}_{-n} - \underline{\mu}\| \geq \varepsilon$.   □

Proof of Theorem 8.3. For $i = 1,\ldots,k$ let

$$SS_i(n) = \sum_{t=1}^{n} 1\{\tau_{i,m} = t, \text{ for some } m\},$$

denote the number of periods in $\{1,\ldots,n\}$ where a forced selection from population i is performed.

Also let,

$$Y_j^b(n) = \sum_{t=1}^{n} 1\{b \in s(\underline{\hat{\mu}}_{-t}), b \text{ is used in period } t, \text{ and } j \text{ is sampled from,}$$

$$\text{due to randomization in } b\}.$$

$$Y^b(n) = \sum_{j \in b} Y_j^b(n),$$

$$Y(n) = \sum_{b \in s(\underline{\mu})} Y^b(n).$$

Since these include all possibilities of selection in a period, it is true that

$$n = \sum_{i=1}^{k} SS_i(n) + \sum_{b \notin s(\underline{\mu})} Y^b(n) + \sum_{b \in s(\underline{\mu})} Y^b(n).$$

Now let W_n denote the sum of outcomes in periods where true optimal BFS are used:

$$W_n = \sum_{b \in s(\underline{\mu})} \sum_{t=1}^{n} X_t \cdot 1\{b \text{ is used in period } t\}.$$

To show the theorem we will prove that

$$\lim_{n \to \infty} \frac{SS_i(n)}{n} = 0, \ i = 1, \ldots, k \tag{8.6}$$

$$\lim_{n \to \infty} \sum_{b \notin s(\underline{\mu})} \frac{Y^b(n)}{n} = 0, \ \text{a.s.,} \tag{8.7}$$

$$\lim_{n \to \infty} \frac{W_n}{n} = z^*(\underline{\mu}), \ \text{a.s..} \tag{8.8}$$

First, (8.6) holds since $\tau_{i,m}$ are sparse for all i. To show (8.7), in no forced selection periods, in order to sample from a BFS b it is necessary but not sufficient that $b \in s(\hat{\underline{\mu}}_n)$, thus

$$Y^b(n) \leq \sum_{t=1}^{n} 1\left\{b \in s(\hat{\underline{\mu}}_t)\right\}.$$

For any $b \in s(\hat{\underline{\mu}}_t)$ and $b \notin s(\underline{\mu})$, it follows from Lemma 8.4 that

$$\|\hat{\underline{\mu}}_t - \underline{\mu}\| \geq \varepsilon.$$

Therefore, for $b \notin s(\underline{\mu})$

$$Y^b(n) \leq \sum_{t=1}^{n} 1\left\{b \in s(\hat{\underline{\mu}}_t)\right\}$$

$$\leq \sum_{t=1}^{n} 1\left\{\|\hat{\underline{\mu}}_t - \underline{\mu}\| \geq \varepsilon\right\}$$

thus,

$$\frac{Y^b(n)}{n} \leq \frac{1}{n}\sum_{t=1}^{n} 1\left\{\|\hat{\underline{\mu}}_t - \underline{\mu}\| \geq \varepsilon\right\} \to 0, \ n \to \infty, \ \text{a.s.,}$$

because $\hat{\underline{\mu}}_t \to \underline{\mu}$, a.s., since $\hat{\underline{\mu}}_t$ is strongly consistent estimator, thus (8.7) holds.

Now to show (8.8) we rewrite W_n as

$$\frac{W_n}{n} = \frac{1}{n} \sum_{b \in s(\underline{\mu})} \sum_{t=1}^{n} X_t \cdot 1\{b \text{ is used in period } t\}$$

$$= \frac{1}{n} \sum_{b \in s(\underline{\mu})} \sum_{j \in b} \sum_{t=1}^{n} X_t \cdot 1\{b \text{ is used in period } t \text{ and } j \text{ is sampled from}\}$$

$$= \frac{1}{n} \sum_{b \in s(\underline{\mu})} \sum_{j \in b} Y_j^b(n) \cdot \overline{X}_{j, Y_j^b(n)}$$

$$= \sum_{b \in s(\underline{\mu})} \frac{Y^b(n)}{n} \cdot \sum_{j \in b} \frac{Y_j^b(n)}{Y^b(n)} \cdot \overline{X}_{j, Y_j^b(n)}.$$

From this expression it follows that

$$\frac{W_n}{n} - z^* = \sum_{b \in s(\underline{\mu})} \frac{Y^b(n)}{n} \cdot \sum_{j \in b} \frac{Y_j^b(n)}{Y^b(n)} \cdot \overline{X}_{j, Y_j^b(n)} - z^*$$

$$= \sum_{b \in s(\underline{\mu})} \frac{Y^b(n)}{n} \cdot z_n^b - z^*,$$

where $z_n^b = \sum_{j \in b} \frac{Y_j^b(n)}{Y^b(n)} \cdot \overline{X}_{j, Y_j^b(n)}.$

Since $Y(n) = \sum_{b \in s(\underline{\mu})} Y^b(n)$, we have

$$\frac{W_n}{n} - z^* = \sum_{b \in s(\underline{\mu})} \frac{Y^b(n)}{n} \cdot z_n^b - z^* + \frac{Y(n)}{n} z^* - \frac{Y(n)}{n} z^*$$

$$= \sum_{b \in s(\underline{\mu})} \frac{Y^b(n)}{n} \cdot (z_n^b - z^*) - (1 - \frac{Y(n)}{n}) z^*.$$

To show (8.8) we will prove that

$$\frac{Y^b(n)}{n} \cdot (z_n^b - z^*) \to 0 \text{ a.s. } \forall b \in s(\underline{\mu}), \text{ and } \frac{Y(n)}{n} \to 1, \text{ a.s.}$$

Random variable $Y^b(n)$ is increasing in n and $0 \le Y^b(n) \le n$, thus either $Y^b(n) \to \infty$ or $Y^b(n) \to M$ for some $M < \infty$. We define the following events:

$$D = \{Y^b(n) \to \infty\} \text{ and } D^c = \{Y^b(n) \to M\}.$$

Now let $P(D) = p$ and $P(D^c) = 1 - p$. Also let

$$A = \left\{ \lim_{n \to \infty} \frac{Y^b(n)}{n} \cdot (z_n^b - z^*) = 0 \right\}.$$

Then $P(A) = P(A|D) \cdot p + P(A|D^c) \cdot (1 - p)$.
  Now,

$$P(A|D) = P\left(\lim_{n \to \infty} \frac{Y^b(n)}{n} \cdot (z_n^b - z^*) = 0 \Big| \lim_{n \to \infty} Y^b(n) = \infty \right)$$

$$\geq P\left(\lim_{n \to \infty} z_n^b - z^* = 0 \Big| \lim_{n \to \infty} Y^b(n) = \infty \right)$$

$$= 1,$$

from the strong law of large numbers, since $\frac{Y^b(n)}{n} \leq 1 \; \forall \, n$, and

$$P(A|D^c) = P\left(\lim_{n \to \infty} \frac{Y^b(n)}{n} \cdot (z_n^b - z^*) = 0 \Big| \lim_{n \to \infty} Y^b(n) = M < \infty \right) = 1,$$

since in this case $z_n^b - z^*$ is bounded for any finite n. Therefore, $P(A) = 1$, thus

$$\frac{Y^b(n)}{n} \cdot (z_n^b - z^*) \to 0, \; n \to \infty, \text{ a.s. }, \; \forall b \in s(\underline{\mu}).$$

Finally,

$$\frac{Y(n)}{n} = \sum_{b \in s(\underline{\mu})} \frac{Y^b(n)}{n} = 1 - \sum_{t=1}^{n} \frac{SS_i(n)}{n} - \sum_{b \notin s(\underline{\mu})} \frac{Y^b(n)}{n} \to 1, \text{ a.s., } n \to \infty.$$

Thus the proof of the theorem is complete.   □

8.4 Rate of Convergence: Simulations

From the results of the previous section it follows that there exists significant flexibility in the construction of a consistent sampling policy. Indeed, any collection of sparse sequences of forced selection periods satisfying (8.3) guarantees that Theorem 8.3 holds.
  In this section we refine the notion of consistency and examine how the rate of convergence of the average outcome to the optimal value is affected by different types of sparse sequences. Furthermore, since the sensitivity analysis will be performed using simulation, it is more appropriate to use the expected value of the deviation as the convergence criterion. We thus consider the expected difference of the average outcome under a consistent policy π from the optimal value:

$$d_n^\pi(\underline{\mu}) = E^\pi \left(\frac{W_n}{n} \right) - z^*(\underline{\mu}).$$

Note that the almost sure convergence of $\frac{W_n}{n}$ to $z^*(\underline{\mu})$ proved in Theorem 8.3 does not imply convergence in expectation, unless further technical assumptions on the unknown distributions are made. For the purpose of our simulation study, we will further assume that the outcomes of any population are absolutely bounded with probability one, i.e., $P(|X_j| \leq u) = 1$, for some $u > 0$. Under this assumption it is easy to show that Theorem 8.3 implies

$$\lim_{n \to \infty} d_n^\pi(\underline{\mu}) = 0, \tag{8.9}$$

for any consistent policy π and any vector $\underline{\mu}$.

To explore the rate of convergence in (8.9), we performed a simulation study, for a problem with $k = 4$ populations. The outcomes of population i follow binomial distribution with parameters (N, p_i), where $p_1 = 0.3, p_2 = 0.5, p_3 = 0.9, p_4 = 0.8$. The vector of expected values is thus $\underline{\mu} = (1.5, 2.5, 4.5, 4)$. The cost vector is $c = (3, 4, 8, 10)$ and $C_0 = 5$. Under this set of values the optimal policy under incomplete information is $x = (0, 3/4, 1/4, 0), y = 0$, and $z^*(\underline{\mu}) = 3$, i.e., it is optimal to randomize between populations 2 and 3, the expected sampling cost per period is equal to 5 and the expected average reward per period is equal to 3.

For the above problem we simulated the performance of a consistent policy for sparse sequences of power function form:

$$\{\tau_{j,m} = \ell_j + m^b, m = 1, 2, \ldots, \}, j = 1, \ldots, k,$$

where ℓ_j are appropriately defined constants which ensure that the sequences are not overlapping, and the exponent parameter b is common for all populations. We compared the convergence rate in (8.9) for five values of b: (1.2, 1.5, 2, 3, 5). For each value of b the corresponding policy was simulated for 1,000 scenarios of length $n = 10^4$ periods each, to obtain an estimate of the expected average outcome per period $d_n^\pi(\underline{\mu})$. The results of the simulations are presented in Fig. 8.1.

We observe in Fig. 8.1 that the convergence is slower both for small and large values of b and faster for intermediate values. Especially for $b = 1.2$ the difference is relatively large even after 10,000 periods. This is explained as follows. For small values of b the forced selections are more frequent. Although this has the desirable effect that the mean estimates for all populations become accurate very soon, it also means that nonoptimal populations are also sampled frequently because of forced selections. As a result the average outcome may deviate from the true optimal value for a longer time period. On the other hand, for large values of b the sequences τ_j all become very sparse and thus the forced selections are rare. In this case it takes a longer time for the estimates to converge, and the linear programming problems may produce nonoptimal solutions for long intervals.

Fig. 8.1 Comparison of
convergence rates for power
sparse sequences

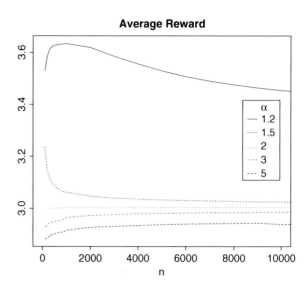

Fig. 8.2 Confidence region for average outcome for $b = 2$

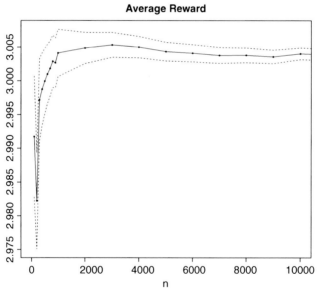

It follows from the above discussion that intermediate values of b are generally preferable, since they offer a better balance of the two effects, fast estimation of all mean values and avoiding nonoptimal populations. This is also evident in the graph, where the value $b = 2$ seems to be the best in terms of speed of convergence.

To address the question of accuracy of the comparison of convergence rates based on simulation, Fig. 8.2 presents a 95% confidence region for the average outcome curve corresponding to $b = 2$, based on 1,000 simulated scenarios. The confidence region is generally very narrow (note that the vertical axes have different scale in the two figures); thus the estimate of the expected average outcome is quite accurate. This is also the case for the other curves; therefore the comparison of convergence rates is valid. Furthermore, the length of the confidence interval becomes smaller for larger time periods since, as expected, the convergence to the true value is better for longer scenario durations.

Another issue arising from Fig. 8.1 is the following. For $b = 1.2$ the average outcome converges very slowly to z^*, but remains above it for the entire scenario duration. Thus it could be argued that, although the convergence is not good, this policy is actually preferable, because it yields higher average outcomes than the other policies. It also seems to contradict the fact that z^* is the maximum average outcome under complete information, since there is a sampling policy that even under incomplete information performs better.

The reason for this discrepancy is related to the form of the cost constraint (8.2). The constraint requires the infinite-horizon expected cost per period not to exceed C_o. This does not preclude the possibility that one or more populations with large sampling costs and large expected outcomes could be used for arbitrarily long intervals before switching to a constrained-optimal policy for the remaining infinite horizon. Such policies might achieve average rewards higher than z^* for long intervals; however this is achieved by "borrowing," i.e., violating the cost constraint, also for long time periods. Since (8.2) is only required to hold in the limit, this behavior of a policy is allowed.

Although the consistent policies in Sect. 8.3 are not designed specifically to take advantage of this observation, they are neither designed to avoid it. Therefore, it is possible, as it happens here for $b = 2$, that a consistent policy may achieve higher than optimal average outcomes for long time periods before it converges to z^*.

The above discussion shows that the constraint as expressed in (8.2), may not be appropriate, if, for example, the sampling cost is a tangible amount that must be paid each time an observation is taken, and there is a budget C_0 per period for sampling. In this situation a policy may suggest exceeding the budget for long time periods and still be feasible, something that may not be viable in reality. In such cases it would be more realistic to impose a stricter average cost constraint, for example, to require that (8.2) holds for all n and not only in the limit.

8.5   Conclusion and Extensions

In this paper we developed a family of consistent adaptive policies for sequentially sampling from k independent populations with unknown distributions under an asymptotic average cost constraint. The main idea in the development of this class of policies is to employ a sparse sequence of forced selection periods for each population, to ensure consistent estimation of all unknown means and in the remaining time periods employ the solution obtained from a linear programming problem that uses the estimates instead of the true values. We also performed a simulation study to compare the convergence rate for different policies in this class.

This work can be extended in several directions. First, as it was shown in Sect. 8.4, the asymptotic form of the cost constraint is in some sense weak, since it allows the average sampling cost to exceed the upper bound for arbitrarily long time periods and still be satisfied in the limit. A more appropriate, albeit more complex, model would be to require the cost constraint to be satisfied at all time points. The construction of consistent and, more importantly, efficient policies under this stricter version of the constraint is work currently in progress.

Another extension is towards the direction of Markov process control. Instead of assuming distinct independent populations with i.i.d. observations, one might consider an average reward Markovian Decision Process with unknown transition law and/or reward distributions, and one or more nonasymptotic side constraints on the average cost. In this case the problem is to construct consistent and, more importantly, efficient control policies, extending the results of [3] in the constrained case.

Acknowledgments This research was supported by the Greek Secretariat of Research and Technology under a Greece/Turkey bilateral research collaboration program. The authors thank Nickos Papadatos and George Afendras for useful discussions on the problem of consistent estimation in a random sequence of random variables.

References

1. Auer, P., Cesa-Bianchi, N., Fischer, P.: Finite-time analysis of the multiarmed bandit. Machine Learning. **47**, 235–256 (2002)
2. Burnetas, A.N., Katehakis, M.N.: Optimal adaptive policies for sequential allocation problems. Adv. App. Math. **17**, 122–142 (1996)
3. Burnetas, A.N., Katehakis, M.N.: Optimal adaptive policies for markovian decision processes. Math. Oper. Res. **22**, 222–255 (1997)
4. Katehakis, M.N., Robbins, H.: Sequential choice from several populations. Proc. Natl. Acad. Sci. USA. **92**, 8584–8585 (1995)
5. Kulkarni, S.R., Lugosi, G.: Finite-time lower bounds for the two-armed bandit problem. IEEE Trans. Automatic Contr. **45**, 711–714 (2000)
6. Lai, T., Robbins, H.: Asymptotically efficient adaptive allocation rules. Adv. App. Math. **6**, 4–22 (1985)
7. Madani, O., Lizotte, D., Greiner, R.: The budgeted multi-armed bandit problem. In: Lecture Notes in Artificial Intelligence, Subseries of Lecture Notes in Computer Science, vol. 3120, pp. 643–645 (2004)
8. Pezeshk, H., Gittins, J.: Sample size determination in clinical trials. Student. **3**(1), 19–26 (1999)
9. Poznyak, A., Nazim, K., Gomez, E.: Self-Learning Control of Finite Markov Chains. CRC Press, New York (2000)
10. Robbins, H.: Some aspects of the sequential design of experiments. Bull. Amer. Math. Monthly. **58**, 527–536 (1952)
11. Wang, Y.G.: Gittins indices and constrained allocation in clinical trials. Biometrika. **78**, 101–111 (1991)

Chapter 9
On a Lanchester Combat Model

G. Kaimakamis and N.B. Zographopoulos

Abstract Theory of combat has to do with the destruction of enemy forces. Most models are descriptive in the sense that they are not built to optimize any particular tactical decision. These models use differential equations and simply describe how the numbers of the opposites forces involved will fluctuate with time, generally decreasing until one or more battle termination criterions are met. After Lanchester, who examined air combat situations during World War I, many researchers studied several models. In this short note we present in brief some of these models and a model using reinforcement.

Keywords Lanchester combat model

Mathematics Subject Classification (2010): 97Mxx

9.1 Introduction

Lanchester in his book Aircraft in Warfare: The Dawn of the Fourth Arm [8] studied several mathematical models which based on differential equations to describe combat situations. Lanchester stated laws on the progression of combat such as the square and linear law. More precisely, Lanchester stated the following model (for two forces—red and blue):

$$\begin{cases} x'(t) = -a\,y(t),\ x(t) > 0, \\ y'(t) = -b\,x(t),\ y(t) > 0, \end{cases}$$

G. Kaimakamis (✉) • N.B. Zographopoulos
Hellenic Army Academy, Department of Mathematics & Engineering Sciences,
16673, Vari, Greece
e-mail: gmiamis@gmail.com; nzograp@gmail.com; zographopoulosn@sse.gr

N.J. Daras (ed.), *Applications of Mathematics and Informatics in Military Science*,
Springer Optimization and Its Applications 71, DOI 10.1007/978-1-4614-4109-0_9,
© Springer Science+Business Media New York 2012

where $x(t)$ and $y(t)$ the number of survivors of blue and red forces at time t and the rate of change of each variable are proportional to the other.

Since then, Lanchester's ideas have been extensively modified from other researchers to study many battles or wars, [2, 10]. More recently MacKay [7] studied the model:

$$
\begin{cases}
x'(t) = -g_1 y_1(t) - g_2 y_2(t), & x(t) > 0, \\
y_1'(t) = -r \frac{y_1(t)}{y(t)} x(t), & y_1(t) > 0, \\
y_2'(t) = -r \frac{y_2(t)}{y(t)} x(t), & y_2(t) > 0,
\end{cases}
$$

where red forces are composed of two types with effectiveness g_1, g_2. Later Koyuncu and Bostanci in [2] focused on the model

$$
\begin{cases}
x'(t) = -a y(t) - b x(t) + x'(t), & x(t) > 0, \\
y'(t) = -c y(t) - d x(t) + y'(t), & y(t) > 0,
\end{cases}
$$

where $-bx(t)$ and $-dy(t)$ define the non-operational loses as a function of the team's own number of units at a given time and $x'(t)$, $y'(t)$ denote the reinforcements.

On the other hand many researchers achieved to verify Lanchester combat models using real data from some famous battles. For example one could see the works of Lucas and Turkes [9] where the battles of Kursk and Ardennes are studied. Also MacKay in [6] deals with the battle of England and Aruka [1] analyzed the battle of Iwo Jima. Finally, we refer to [3,5], where the interested reader may found applications of the Lanchester combat models to insurgency situations.

9.2   A Model with Reinforcement

In this note we consider a Lanchester Combat Model with reinforcement. We consider the system of equations

$$
1 \begin{cases}
x'(t) = -a y(t) + g(t - \tau) y(t), \\
y'(t) = -b x(t),
\end{cases}
\tag{9.1}
$$

where

$$
g(t - \tau) = \begin{cases}
0, & t < \tau, \\
g_0, & t \geq \tau,
\end{cases}
$$

g_0 be a positive constant number. The initial conditions are denoted by $x_0 > 0$ and $y_0 > 0$, respectively. For the positive numbers a, b, we assume that

$$
b x_0^2 < a y_0^2.
\tag{9.2}
$$

We try to give the optimal reinforcement of the army x and the corresponding time τ, such that x wins.

9.2.1 For $t < \tau$

When $t < \tau$ system (9.1) has the form

$$\begin{cases} x'(t) = -ay(t), \\ y'(t) = -bx(t). \end{cases} \tag{9.3}$$

The dynamics of system (9.3) are given by standard methods. More precisely, for any $t < \tau$ the following equality holds:

$$-bx^2(t) + ay^2(t) = C,$$

where C is given by

$$-bx_0^2 + ay_0^2 = C. \tag{9.4}$$

Note that $x(t)$ and $y(t)$ are both decreasing functions of time. Our assumption (9.2) implies that, for large t, x will be eliminated or that y finally wins. However, this is not the case we discuss here. Thus the reinforcement of the army x must happen before its elimination.

We calculate the precise value of the time of this elimination; the solution of the system (9.3) is given by

$$\begin{aligned} x(t) &= c_1 \sqrt{a}\,e^{-\sqrt{ab}\,t} + c_2 \sqrt{a}\,e^{\sqrt{ab}\,t}, \\ y(t) &= c_1 \sqrt{b}\,e^{-\sqrt{ab}\,t} - c_2 \sqrt{b}\,e^{\sqrt{ab}\,t}, \end{aligned} \tag{9.5}$$

where

$$c_1 = \frac{1}{2}\left(\frac{x_0}{\sqrt{a}} + \frac{y_0}{\sqrt{b}} \right) \quad \text{and} \quad c_2 = \frac{1}{2}\left(\frac{x_0}{\sqrt{a}} - \frac{y_0}{\sqrt{b}} \right).$$

Observe that c_2 is a negative quantity. From these relations we calculate that x becomes zero at the time

$$\tau_{\text{elim}} = \frac{1}{2\sqrt{ab}} \log\left(\frac{c_1}{-c_2} \right). \tag{9.6}$$

We thus obtain that an upper bound for τ is τ_{elim}:

$$\tau < \tau_{\text{elim}}. \tag{9.7}$$

9.3   For $t > \tau$

When $t > \tau$ system (9.1) has the form

$$\begin{cases} x'(t) = -ay(t) + g_0 y(t), \\ y'(t) = -bx(t). \end{cases} \tag{9.8}$$

The dynamics of system (9.3) is described by the following equality:

$$-bx^2(t) + ay^2(t) - g_0 y^2(t) = C_\tau,$$

where C_τ is given by

$$-bx^2(\tau) + ay^2(\tau) - g_0 y^2(\tau) = C_\tau. \tag{9.9}$$

The continuity of the solutions of (9.1) implies that $x(\tau)$ and $y(\tau)$ are given by (9.5). The army x wins if the constant C_τ is negative. This means that

$$-bx^2(\tau) + ay^2(\tau) < g_0 y^2(\tau),$$

or

$$g_0 > \frac{C}{y^2(\tau)}, \tag{9.10}$$

where C is given by (9.4). From (9.10) we finally conclude that the optimal reinforcement of the army x at the time τ is

$$g_0 y(\tau) = \frac{C}{y(\tau)}, \tag{9.11}$$

and the optimal time for this is as $t \to 0$, i.e., as sooner as possible. In any case (9.7) must hold.

9.4   Conclusion

Summarizing we conclude that the best time for the reinforcement is as much as earlier, with the restriction that this will happen before the elimination time τ_{elim} given in (9.6). The number of this reinforcement is given in terms of the enemy $y(t)$, by (9.10).

References

1. Aruka, Y.: Some adaptive economic processes in social interactions. IBRCU Working Paper No. 9 (2003)
2. Isac, G., Gosselin, A.: A military application of viability: winning cones, differential inclusions, and Lanchester type models for combat. In: Chinchuluun, A. Pardalos, P.M., Migdalas, A., Pitsoulis, L. (eds.) Pareto Optimality, Game Theory And Equilibria, Springer, Berlin (2008)
3. Kaplan, E.H., Kress, M. Szechtman, R.: Confronting entrenched insurgents. Oper. Res. **58**(2), 329–341 (2010)
4. Koyuncu, B., Bostanci, E.: Using Lancaster combat models to aid battlefield visualization, 2nd IEEE International Conference on Computer Science and Information Technology. ICCSIT, China, 290–292 (2009)
5. Kress, M., Szechtman, R.: Why defeating insurgencies is hard: the effect of intelligence in counterinsurgency operations–a best-case scenario. Oper. Res. **57**(3), 578–585 (2009)
6. MacKay, N.J.: Mathematical models of the battle of Britain, IMA Conference, Mathematics in Defence (2009)
7. MacKay, N.J.: Lanchester combat models, arxiv.0606300 (2006)
8. Lanchester, F.W.: Aircraft in Warfare: The Dawn of the Fourth Arm. Constable and Co. Ltd., London, England (1916)
9. Lucas, T.W., Turkes, T.: Fitting Lanchester equations to the battles of Kursk and Ardennes. Nav. Res. Logist. **51**(1), 95–116 (2004). doi: 10.1002/nav.10101
10. Wiper, M.P., et al.: Bayesian inference for a Lanchester type combat model. Nav. Res. Logist. **47**(7), 541–558 (2000)

Chapter 10
Land Warfare and Complexity

Dionysios Stromatias

Abstract This issue summarizes the results of a multiyear research program whose basic chapter was to use complex adaptive systems theory to develop tools to help to understand the fundamental processes of war. The chapters are mostly self-contained, so that they may be read in any order, and are roughly divided into two parts. Part one introduces the general context for the ensuing discussion, and provides both qualitative and more technical overviews of those elements of nonlinear dynamics, artificial-life, complexity theory and multiagent-based simulation tools that are applied to modeling combat. Part two summarizes the main ideas introduced throughout the issue.

Keywords Complex systems • Self-organization • Collectivism • Edge of chaos • Cellular automata • Genetic algorithms • Levels of applicability of complexity theory

10.1 Introduction

In 1914, the British engineer F.W. Lanchester developed a theory based on World War I aircraft engagements to explain why concentration of forces was useful in modern warfare. His model underlies many low-resolution and medium-resolution combat models and works on the basis of attrition

1. Homogeneous (model)

   (a) A single scalar represents a unit's combat power.
   (b) Both sides are considered to have the same weapon effectiveness.

D. Stromatias
Hellenic Artillery School, Nea Peramos Attikis, Greece
e-mail: iamstromi@yahoo.gr

N.J. Daras (ed.), *Applications of Mathematics and Informatics in Military Science*,   119
Springer Optimization and Its Applications 71, DOI 10.1007/978-1-4614-4109-0_10,
© Springer Science+Business Media New York 2012

2. Heterogeneous (model)

   (a) Attrition is assessed by weapon type and target type and other variability
       factors.

   Similar forms also apply to models of biological populations in ecology.
   Lanchester (deterministic) equations describe the rate at which a force loses
systems as a function of the size x of the force and the size of the enemy force y

$$\frac{dx}{dt} = f_1(x, y, \dots)$$

$$\frac{dy}{dt} = f_2(x, y, \dots).$$

Solving these equations as functions of $x(t)$ and $y(t)$ provide insights about
battle outcome. The most elementary form arises by taking $f_1(x, y, \dots) = -ay$
and $f_2(x, y, \dots) = -bx$:

$$\frac{dx}{dt} = -ay$$

$$\frac{dy}{dt} = -bx.$$

Integrating, the equations which describe modern warfare we get the following state
equation, called Lanchester's "Square Law":

$$b(x_0^2 - x^2) = a(y_0^2 - y^2).$$

These equations have also been postulated to describe "aimed fire" where $\sqrt{ab}$
measures *battle intensity* and $\sqrt{a/b}$ measures *relative effectiveness*.
After extensive derivation, the following expression for the X force level is derived
as a function of time (the Y force level is equivalent):

$$x(t) = \frac{1}{2}\left[\left(x_0 - \sqrt{\frac{a}{b}}y_0\right)e^{\sqrt{ab}t} + \left(x_0 + \sqrt{\frac{a}{b}}y_0\right)e^{-\sqrt{ab}t}\right].$$

To determine who will win, each side must have victory conditions, i.e., we must
have a "battle termination model." Assume both sides fight to annihilation. One of
three outcomes at time t_f, the end time of the battle:

1. X wins, i.e., $x(t_f) > 0$ and $y(t_f) = 0$.
2. Y wins, i.e., $y(t_f) > 0$ and $x(t_f) = 0$.
3. Draw, i.e., $x(t_f) = 0$ and $y(t_f) = 0$.

It can be shown that a Square-Law battle will be won by X if and only if:

$$\frac{x_0}{y_0} > \frac{a}{b}.$$

Now, several natural questions arise. How many survivors x_f are there when X wins a fight-to-the-finish? And, when X wins, how long does it take? The answer is an easy consequence of the above formulas. Indeed, we have

$$x_f = \sqrt{\left(x_0^2 - \frac{a}{b}y_0^2\right)}.$$

The time needed is

$$t\left(x_f\right) = \frac{1}{2\sqrt{ab}} \ln \left[\frac{1 + \frac{y_0}{x_0}\sqrt{\frac{a}{b}}}{1 - \frac{y_0}{x_0}\sqrt{\frac{a}{b}}}\right].$$

Further, assuming battle termination at $x(t) = x_{BP}$ or $y(t) = y_{BP}$, in what case X wins and how long does it take if X wins? It can be shown that X wins if and only if

$$\frac{x_0}{y_0} > \sqrt{\frac{a}{b}\left(\frac{1 - [y_{BP}^2/y_0^2]}{1 - [x_{BP}^2/x_0^2]}\right)}.$$

In this case, the time needed for X's victory is given by

$$t\left(y_{BP}\right) = \begin{cases} \frac{1}{\sqrt{ab}}\ln\frac{x_0}{x_{BP}}, \; if \; \frac{x_0}{y_0} = \sqrt{\frac{a}{b}} \\ \frac{1}{\sqrt{ab}}\ln\left(\frac{y_{BP} - \sqrt{y_{BP}^2 - y_0^2 + (b/a)x_0^2}}{y_0 - \sqrt{(b/a)}x_0}\right), \; if \; \frac{x_0}{y_0} \neq \sqrt{\frac{a}{b}}. \end{cases}$$

10.2 Complex Adaptive Systems

The main idea put forth in this paper is that significant new insights into the fundamental processes of land warfare can be obtained by viewing land warfare as a *complex adaptive system* (CAS). That is to say, by viewing a military "conflict" as a nonlinear dynamical system composed of many interacting semiautonomous and hierarchically organized agents continuously adapting to a changing environment [2].

Complex systems is a new field of science studying how parts of a system give rise to the collective behaviors of the system, and how the system interacts with its environment. Social systems formed (in part) out of people, the brain formed out of neurons, molecules formed out of atoms, the weather formed out of air flows are all examples of complex systems. The field of complex systems cuts across all traditional disciplines of science, as well as engineering, management, and medicine. It focuses on certain questions about parts, wholes, and relationships. These questions are relevant to all traditional fields.

The study of complex systems is about understanding indirect effects. Problems that are difficult to solve are often hard to understand because the causes and effects are not obviously related. Pushing on a complex system "here" often has effects "over there" because the parts are interdependent. This has become more and more apparent in our efforts to solve societal problems or avoid ecological disasters caused by our own actions. The field of complex systems provides a number of sophisticated tools, some of them are concepts that help us think about these systems, some of them are analytical for studying these systems in greater depth, and some of them are computer based for describing, modeling, or simulating these systems.

There are three interrelated approaches to the modern study of complex systems

1. How interactions give rise to patterns of behavior
2. Understanding the ways of describing complex systems
3. The process of formation of complex systems through pattern formation and evolution

CST is concerned with more complicated systems, where "complicated" typically means that a system consists of a large number of mutually interrelated parts. In dealing with such systems, CST generalizes the conventional approach in two fundamental ways:

1. The final state, S_{final}, is no longer assumed to be a function of the initial state alone, but can depend strongly on the *path*, P, that the system follows in evolving from its initial to final states.
2. The initial state is endowed with both an *internal* and *external* structure. CST can be described as *the study of the behavior of collections of simple (and typically nonlinearly) interacting parts that can evolve*.

10.3 Military Conflicts, Particularly Land Combat, Have Almost All the Key Features of Complex Adaptive Systems

Nonlinear interaction

- Friendly and enemy forces are composed of a large number of nonlinearly interacting "parts."
- Combat is not just an aggregate of many smaller-scale conflicts, but is a complex system composed of parts whose action and pattern of behavior depend on the action and pattern of behavior of other (nearby and not-so-nearby) parts.

Decentralized control

- Despite the presence of "global commanders," who have a (global, albeit imprecise) view of the overall combat arena, there is no master "voice" that dictates the actions of each and every combatant.

Self-organization

- Local action, which often appears "chaotic," induces long-range order.
- Command and control tends to organize what is otherwise disorganized action.

Collectivism

- There is continual feedback between the behavior of (low-level) combatants and the (high-level) C2 hierarchy.

Parts are more like "niches" than "parts"

- Their parts, particularly those represented by the lowest level combatant and as long as the war fighting skills of combatants exceed some threshold war fighting skill level, are essentially interchangeable.

Adaptation

- Their parts, in order to survive, must continually adapt to a changing combat environment (new strategies and tactics must be conceived of and implemented on-the-spot and in immediate response to changes in the environment).
- Each combatant comes into a conflict armed with a set of default rules ("doctrine"), a goal (or goals), and hardware designed to facilitate the implementation of doctrine. The success or failure of a campaign depends on how well each combatant adapts to the continually changing combat environment, which includes the functioning and adaptation of both friendly and enemy combatants.
- Actions and outcomes of actions are as much a function of the internal "human element" (reasoning capacity, unpredictability, inspiration, accident, etc.) as they are of the hardware.

Hierarchical structure

- Parts are organized in a (command and control) hierarchy.

10.3.1 Self-Organization

Self-organization is a fundamental characteristic of complex systems. It refers to the emergence of macroscopic nonequilibrium organized structures, and is due to the collective interactions of the constituents of a complex system as they react and adapt to their environment.

There is no God-like "oracle" dictating what each and every part ought to be doing; parts act locally on local information and global order emerges without any need for external control.

10.3.2   Collectivism

As an example of the importance of collectivism, consider a natural ecology. Each species that makes up an ecology composed of a large number of diverse species coevolves with other members of the ecology according to a fitness function that is, in part, itself a function of the emerging ecology. Individual members of each species collectively define a (part of the) coevolving ecology; the ecology, in turn, determines the fitness-function according to which its constituent parts evolve. It is this nonlinear feedback between the information describing individual species (or the system's microscopic level) and the global ecology (or the system's macroscopic level) that those species collectively define that determines the temporal evolution—and identity—of the entire system.

10.3.3   Chaos and Complexity

Very loosely speaking, it can be said that where *chaos* is the study of how simple systems can generate complicated behavior, *complexity* is the study of how complicated systems can generate simple behavior. Since both chaos and complex systems theory attempt to describe the behavior of dynamical systems, it should not be surprising to learn that both share many of the same tools, although, properly speaking, complex systems theory ought to be regarded as the superset of the two methodologies.

10.4   Cellular Automata

A cellular automaton is a collection of "colored" cells on a grid of specified shape that evolves through a number of discrete time steps according to a set of rules based on the states of neighboring cells. The rules are then applied iteratively for as many time steps as desired. Von Neumann was one of the first people to consider such a model, and incorporated a cellular model into his "universal constructor." Cellular automata were studied in the early 1950s as a possible model for biological systems. Comprehensive studies of cellular automata have been performed by S. Wolfram starting in the 1980s, and Wolfram's fundamental research in the field culminated in the publication of his book *A New Kind of Science* in which Wolfram presents a gigantic collection of results concerning automata, among which are a number of ground breaking new discoveries. Cellular automata come in a variety of shapes and varieties. One of the most fundamental properties of a cellular automaton is the type of grid on which it is computed. The simplest such "grid" is a one-dimensional line. In two dimensions, square, triangular, and hexagonal grids may be considered. Cellular automata may also be constructed on Cartesian grids in arbitrary numbers of dimensions, with the d-dimensional integer lattice z^d being the most common choice. Cellular automata on a d-dimensional integer lattice are implemented in Mathematica as Cellular Automaton [*rule, init, steps*].

The number of colors (or distinct states) k a cellular automaton may assume must also be specified. This number is typically an integer, with $k = 2$ (binary) being the simplest choice. For a binary automaton, color 0 is commonly called "white," and color 1 is commonly called "black". However, cellular automata having a continuous range of possible values may also be considered.

In addition to the grid on which a cellular automaton lives and the colors its cells may assume, the neighborhood over which cells affect one another must also be specified. The simplest choice is "nearest neighbors," in which only cells directly adjacent to a given cell may be affected at each time step. Two common neighborhoods in the case of a two-dimensional cellular automaton on a square grid are the so-called Moore neighborhood (a square neighborhood) and the von Neumann neighborhood (a diamond-shaped neighborhood).

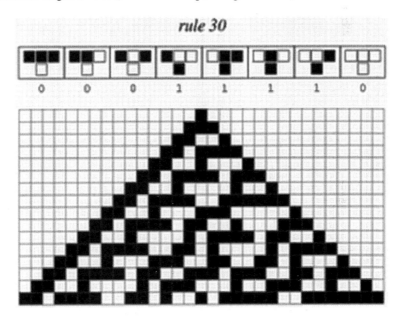

The simplest type of cellular automaton is a binary, nearest-neighbor, one-dimensional automaton. Such automata were called "elementary cellular automata" by S. Wolfram, who has extensively studied their amazing properties. There are 256 such automata, each of which can be indexed by a unique binary number whose decimal representation is known as the "rule" for the particular automaton. An illustration of rule 30 is shown above together with the evolution it produces after 15 steps starting from a single black cell.

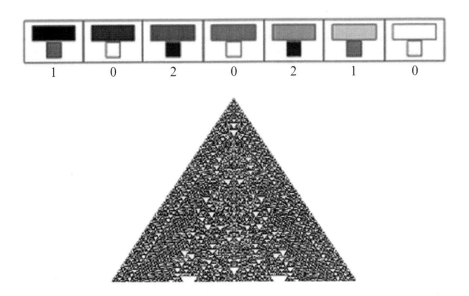

A slightly more complicated class of cellular automata are the nearest-neighbor, k-color, one-dimensional totalistic cellular automata. In such automata, it is the *average* of adjacent cells that determine the evolution, and the simplest nontrivial examples have $k = 3$ colors. For these automata, the set of rules describing the behavior can be encoded as a $(3k - 2)$-digit k-ary number known as a "code." The rules and 300 steps of the ternary ($k = 3$) code 912 automaton are illustrated above.

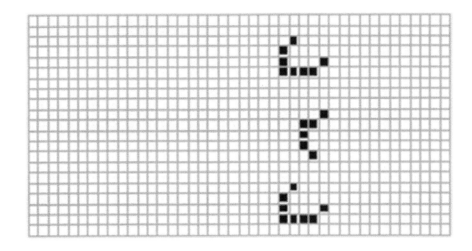

In two dimensions, the best-known cellular automaton is Conway's game of life, discovered by J.H. Conway in 1970 and popularized in Martin Gardner's *Scientific American* columns. The game of life is a binary ($k = 2$) totalistic cellular automaton with a Moore neighborhood of range $r = 1$. Although the computation of successive game of life generations was originally done by hand, the computer revolution soon arrived and allowed more extensive patterns to be studied and propagated. An animation of the game of life construction known as a puffer train is illustrated above. The theory of cellular automata is immensely rich, with simple rules and structures being capable of producing a great variety of unexpected behaviors. For example, there exist universal cellular automata that are capable of simulating the behavior of any other cellular automaton or Turing machine. It has even been proved by Gacs (2001) that there exist fault-tolerant universal cellular automata, whose ability to simulate other cellular automata is not hindered by random perturbations provided that such perturbations are sufficiently sparse.

10.5 Genetic Algorithms

Genetic algorithms are one of the best ways to solve a problem for which little is known. They are very general algorithms and so will work well in any search space. All you need to know is what is required for the solution to be able to do well, and a genetic algorithm will be able to create a high-quality solution. Genetic algorithms use the principles of selection and evolution to produce several solutions to a given problem. Genetic algorithms tend to thrive in an environment in which there is a very large set of candidate solutions and in which the search space is uneven and has many hills and valleys. True, genetic algorithms will do well in any environment, but they will be greatly outclassed by more situation-specific algorithms in the simpler search spaces. Therefore you must keep in mind that genetic algorithms are not always the best choice. Sometimes they can take quite a while to run and are therefore not always feasible for real-time use. They are, however, one of the most powerful methods with which to (relatively) quickly create high-quality solutions to a problem. The most common type of genetic algorithm works like this: a population is created with a group of individuals created randomly. The individuals in the population are then evaluated. The evaluation function is provided by the programmer and gives the individuals a score based on how well they perform at the given task. Two individuals are then selected based on their fitness, the higher the fitness, the higher the chance of being selected. These individuals then "reproduce" to create one or more offspring, after which the offspring are mutated randomly. This continues until a suitable solution has been found or a certain number of generations have passed, depending on the needs of the programmer.

10.5.1 Swarms

Swarm intelligence is the collective behavior of decentralized, self-organized systems, natural or artificial. The concept is employed in work on artificial intelligence. The expression was introduced by Gerardo Beni and Jing Wang in 1989, in the context of cellular robotic systems [1]. Swarm intelligence systems are typically made up of a population of simple agents or boids, interacting locally with one another and with their environment. The agents follow very simple rules, and although there is no centralized control structure dictating how individual agents should behave, local and to a certain degree random interactions between such agents lead to the emergence of intelligent global behavior, unknown to the individual agents. Swarm intelligence research is multidisciplinary. It can be divided into natural swarm research studying biological systems and artificial swarm research studying human artifacts. There is also a scientific stream attempting to model the swarm systems themselves and understand their underlying mechanisms, and an engineering stream focused on applying the insights developed by the scientific stream to solve practical problems. The goal of the Swarm project is to provide the complex systems theory research community with a fully general-purpose artificial-life simulator. The system comes with a variety of generic artificial worlds populated with generic agents, a large library of design and analysis tools and a "kernel" to drive the actual simulation. These artificial worlds can vary widely, from simple 2D worlds in which elementary agents move back and forth to complex multidimensional "graphs" representing multidimensional telecommunication networks in which agents can trade messages and commodities, to models of real-world ecologies in other areas.

10.6   Complex Adaptive Systems

CASs are complex systems (meaning that they consist of many nonlinearly interacting parts) whose parts can adapt to changing environments. Moreover, each "part" typically exists within a nested hierarchy of parts within parts.

Traditionally, simulations of complex systems have consisted of mathematical or stochastic models, typically involving differential equations that relate one set of global parameters to another set and describe the system's overall dynamics. The behavior of a system is then "understood" by looking at the relationship between the input and output variables of the simulation. While such an approach is adequate for systems with parts that possess little or no internal structure, it is largely incapable of describing groups, or societies, in which the *internal dynamics* of the constituent members of the system represent a vital part of the underlying dynamics.

Additional drawbacks of traditional simulation methods include:

- A failure to distinguish among *different levels of activity* within real complex systems; that is to say, a failure to appreciate that global parameters, such as the population size of an ecology, are often profoundly related to local parameters, such as the decision-making processes of individuals within the ecology—traditional simulation methods, particularly those relying on a differential equation approach, seldom take into account this local-global dichotomy

- An inability to *analytically* account (such as in a differential equation form) for individual actions and/or strategies of the constituent elements of a complex system
- An inability to realistically account for the *qualitative information* that individuals may use in formulating their strategies and upon which they may base their local decision

The fundamental question that is addressed, at least indirectly, in this report, and more fully in the follow-on paper, is

"what does complexity theory tell us about land warfare?"

This question really embodies three separate but interrelated issues:

1. *Complexity theory*
2. *Land warfare*
3. *Modeling/simulation*

The figure above shows that there are four levels of applicability of complexity theory:

- *Level-1*, consisting of specific analytical and mathematical tools such as cellular automata, genetic algorithms, genetic programming, and so on
- *Level-2*, consisting of general simulation systems such as SWARM, within which complex systems can be modeled
- *Level-3*, consisting of observations of behavior of specific systems
- *Level-4*, consisting of sets of universal behaviors, such as the principle of self-organized criticality

Ideally, of course, one would like to take whatever insights complexity theory has come up with, or will come with, on the highest level (*level-4*) and apply them directly to the issues and problems of land warfare. The fact that this is exceedingly unlikely to happen in the foreseeable future is due in no small measure to the fact that, as of this writing, there are precious few "universal behaviors" populating *level-4*.

Indeed, as alluded to in an earlier section, self-organized criticality is arguably the *only* existing holistic mathematical theory of self-organization in complex systems!

Therefore, if there is anything at all that falls under the rubric of complexity theory that is generally applicable to the problems of land warfare, it will most likely consist of specific sets of tools applied to specific problems, along with whatever insights can be gained by using general-purpose simulators such as SWARM to act as simulation "engines." There remains the possibility that complexity theory might shed some light on how battlefields may be configured (or compelled to self-organize) to achieve a maximum adaptability to a changing environment.

The figure also shows that there are four levels of land warfare to which the tools and methodologies of complexity theory can be applied:

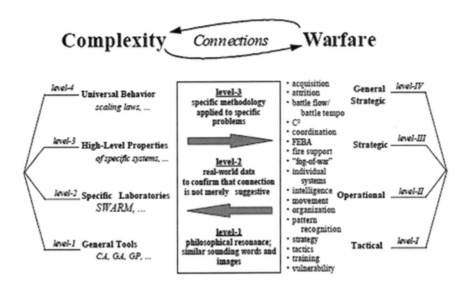

1. *Tactical*
2. *Operational*
3. *Strategic*
4. *General strategic*, which refers to the sociopolitical strategies that are followed over long periods of time and which can therefore span over several conflicts

Finally, the figure illustrates that there are three levels on which complexity theory can be applied to land warfare:

• *Level-1*. This is the most basic *metaphor level* to which most general discussions have been heretofore confined. This level consists of constructing and elaborating upon similar sounding words and images that most strongly suggest a "philosophical resonance" between behaviors of complex systems and certain aspects of what happens on a battlefield. The Clauswitzian images of "fog of war" and "friction" come to mind immediately. There is nothing wrong, per se, with confining a discussion to this level, but one must always be mindful of the fact that metaphors are easily abused and "philosophical resonances" do not imply real connections.

• *Level-2*. This is the pragmatic and/or experimental level on which real-world data are mined to confirm or deny that there is more to a possible connection between complexity theory and land warfare than mere "philosophical resonance" alone. The best work along these lines has so far been conducted by Tagarev and is discussed briefly below. Tagarev provides evidence of deterministic chaos in tactical, operational, and strategic dynamics of a wide class of military behavior.

• *Level-3*. This is the "workhorse" level on which specific methodology borrowed from complexity theory is applied directly to specific issues and problems of land warfare. This might not be as intellectually provocative or satisfying as making a direct, one-to-one mapping between universal patterns of behavior of complex systems in general and patterns of combat on the battlefield (although this is remotely conceivable in some form); however, using genetic algorithms to evolve tactics in real-time in the heat-of-battle is impressive nonetheless. Most of the ideas and conjectures outlined in the following sections fall squarely into this third level of connections.

10.7 Conclusions

This report concludes that the concepts, ideas, theories, tools, and general methodologies of nonlinear dynamics and complex systems theory show enormous, almost unlimited, potential for not just providing better solutions for certain existing problems of land combat, but for *fundamentally altering our general understanding of the basic processes of war, at all levels.* Indeed, the new sciences' greatest legacy may, in the end, prove to be not just a set of creative answers to old questions but an *entirely new set of questions* to be asked of what really happens on the battlefield.

The central idea of this paper is that *land combat is a complex adaptive system.* That is to say that land combat is essentially a nonlinear dynamical system composed of many interacting semiautonomous and hierarchically organized agents continuously adapting to a changing environment.

Military conflicts, particularly land combat, have almost all of the key features of CAS:

• Combat forces are composed of large numbers of nonlinearly interacting parts and are organized in a command and control hierarchy; local action, which often appears disordered, induces long-range order (i.e., combat is self-organized); military conflicts, by their nature, proceed far from equilibrium; military forces, in order to survive, must continually adapt to a changing combat environment; there is no master "voice" that dictates the actions of each and every combatant (i.e., battlefield action effectively proceeds according to a decentralized control); and so on. In principle, this means that land combat ought to be amenable to precisely the same methodological course of study as any other complex adaptive system, such as the stock market, a natural ecology or the human brain.

Implicitly in this paper is the idea that these largely conceptual links between properties of land warfare and properties of complex systems in general can be extended to forge a set of practical connections as well. That is to say, land warfare does not just *look like* a complex system on paper, but can be well characterized in practice using the same basic principles that are used for discovering and identifying behaviors in complex systems.

References

1. Beni, G., Wang, J.: Intelligence in cellular robotic systems. Proceedings of NATO Advanced Workshop on Robots and Biological Systems (1989)
2. Ilachinski, A.: Land warfare and complexity, Part I: mathematical background and technical sourcebook. In: book or Journal, pages 124–125

Part VI
Satellite Remote Sensing

Chapter 11
Wavelet Transform in Remote Sensing with Implementation in Edge Detection and Noise Reduction

Pantelis N. Michalis

Abstract In image processing, the Fourier transform has a serious drawback as only frequency information remains whilst local information is lost. In order to involve localization in the analysis, the Short Time Fourier Transform (STFT) is adapted where the image is windowed. The drawback is that the window is the same in all frequencies. In principle, a more flexible approach is required where the window size varies in order to determine more precisely either location or frequency. Wavelet analysis allows the variation of the window based on the frequency information. Wavelets have limited duration and an average value of zero and thus they are irregular and asymmetric with short duration. Wavelets can be used in the field of edge detection and enhancement, image compression, noise reduction, and image fusion. In this review paper wavelets are used in quite opposite applications such as edge detection and noise reduction of remote sensing images. Thus, the flexibility and versatility of the wavelets is exposed. The challenge is to choose the appropriate wavelet for a particular application which is not known a priori.

Keywords Wavelets • Remote sensing • Multiresolution • Scale • Edge detection • Noise reduction

Mathematics Subject Classification (2010): 65T50, 65T60

P.N. Michalis
Greek Ministry of Defence, Mesogeion 227-231, Athens
e-mail: pantelis.michalis@gmail.com

N.J. Daras (ed.), *Applications of Mathematics and Informatics in Military Science*,
Springer Optimization and Its Applications 71, DOI 10.1007/978-1-4614-4109-0_11,
© Springer Science+Business Media New York 2012

11.1 Introduction

The Fourier transform has been the mainstay of transform-based image processing since the late 1950s. However, Fourier analysis has a serious drawback as only frequency information remains while the local one is lost. This means that any modification of the Fourier coefficients has a global effect on the image. In order to involve localization on the analysis, the Short Time Fourier Transform (STFT) is adapted [1]. In this case, the image is windowed, and thus the information has a precision relevant to the size of the window used. The drawback is that the window is the same in all frequencies. In principle, a more flexible approach is required in image processing, where the window size varies in order to determine more precisely either location or frequency.

Wavelet analysis allows the variation of the window based on the frequency information. This means that long time intervals are used in low-frequency information and short time intervals in high-frequency information. In general, wavelets have limited duration and an average value of zero and they are irregular and asymmetric with short duration [4].

There are many possible sets of wavelets which are represented by a mother wavelet and a scaling function. The challenge is to choose the appropriate wavelet for a particular application something which is not known a priori. Various wavelets have been introduced along with the suitable threshold level for the adjustment of the wavelet coefficients, attempting to give better results in various applications.

In this paper this versatility of the wavelets is shown in quite contradictory applications, the edge detection on one hand and the noise reduction on the other.

11.2 Wavelets

As with the Fourier transform, the same possibilities exist for wavelet transforms: a continuous wavelet transform (CWT), a wavelet series expansion, and a discrete wavelet transform (DWT).

11.2.1 The Continuous Wavelet Transform (CWT)-Wavelet Properties

11.2.1.1 Introduction

Digital images are discrete as series of pixels in two dimensions. However, for completeness, the CWT is covered in this paragraph giving a direct comparison to the Fourier transform [1].

The CWT of a continuous, square-integrable function, $f(t)$ is described as

$$\gamma(s,t) = \int f(t)\psi_{s,t}^*(t)dt, \tag{11.1}$$

where $*$ denotes complex conjugation. This equation shows how a function $f(t)$ is decomposed into a set of wavelets $\psi_{s,\tau}(t)$. The variables s and t are the scale and the translation. The wavelets are generated from a single basic *mother* wavelet $\psi(t)$, by scaling and translation:

$$\psi_{s,t}(t) = \frac{1}{\sqrt{s}}\psi\left(\frac{t-\tau}{s}\right). \tag{11.2}$$

The theory of wavelet transforms defines a framework within which a wavelet is designed in order to fulfil specific criteria. It is not needed to specify the wavelet basis functions a priori. This is another advantage of the wavelet transform compared to the Fourier transform, or other transforms.

The most important properties of wavelets are the admissibility and the regularity conditions. Wavelets should be waves as they are used to analyze and reconstruct a signal without loss of information (Admissibility condition).

An additional condition on the wavelet functions should be defined in order to make the wavelet transform decrease quickly with scale s. This is the *regularity condition* where the wavelets should have some smoothness and concentration in both time and frequency domains. Regularity is explained using the concept of *vanishing moments*.

To summarize, the "wave" is given by the admissibility condition, while the "let" or fast decay by the regularity condition, and thus the wavelet is developed [2].

11.2.1.2 The Two-Dimensional CWT

As can be seen from equation (11.1) the wavelet transform of a one-dimensional function $f(t)$ is two-dimensional. For functions of more than one variable, this transform also increases the dimensionality by one.

11.2.2 The Discrete Wavelet Transform

11.2.2.1 Introduction

When digital images are to be viewed or processed at multiple resolutions, the *DWT* is the mathematical tool of choice [2]. In addition to being an efficient, highly intuitive framework for the representation and storage of multiresolution images, the DWT provides powerful insight into image's spatial and frequency characteristics. An efficient way to implement DWT using filters was developed by Mallat [4]. The Mallat algorithm is in fact a classical scheme known in the signal-processing

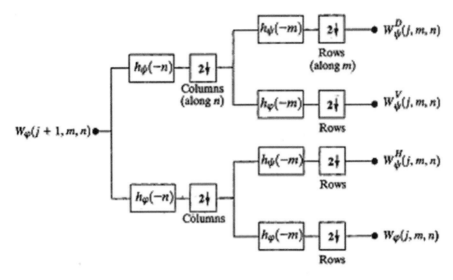

Fig. 11.1 Fast wavelet transform (source [2])

community as a *two-channel subband coder*. This very practical filtering algorithm yields a *fast wavelet transform*—a box into which a signal passes, and out of which wavelet coefficients quickly emerge.

11.2.2.2 Fast Wavelet Transform (FWT)

An important consequence of the above properties is that both $\phi(x)$ and $\psi(x)$ can be expressed as linear combinations of double-resolution copies of themselves. That is, via the series expansions

$$\varphi(x) = \sum h_\varphi(n)\sqrt{2}\varphi(2x - n)$$
$$\psi(x) = \sum h_\psi(n)\sqrt{2}\psi(2x - n),$$

where h_ϕ and h_ψ are called *scaling and wavelet vectors*, respectively. They are the filter coefficients of the FWT, an iterative computational approach to the DWT shown in Fig. 11.1. The $W_\phi(j,m,n)$ and $W_\psi^i(j,m,n)$ for H,V,D outputs in Fig. 11.1 are the DWT coefficients at scale j. Blocks containing time-reserved scaling and wavelet vectors are the low-pass and high-pass decomposition. Finally, blocks containing a 2 and down arrow represent downsampling.

Each pass thought the filter bank decomposes the input into four lower resolution components. The W coefficients are created via two low-pass filters and thus called approximation coefficients. $W_\psi^i(j,m,n)$ for H,V,D} are horizontal, vertical, and diagonal coefficients, respectively. This is the representation of the first iteration (Fig. 11.1). The second iteration would consider as input the approximation coefficients.

11.3   Wavelets in Edge Detection and Noise Reduction

The specifications (resolution, color depth, and geo-location accuracy) of high-resolution satellite images which have been acquired since 1999 are directly comparable to airphotos specifications, gives a boost in Remote Sensing to expand in new fields of applications. In such fields, edges are of great importance as they determine the boundaries of significant objects like buildings, urban vegetation, and road networks, river banks, flood influence, etc. However, the key factor is to handle the image noise in such a way that it is not represented as an edge.

The edge detection problem has traditionally been addressed with the use of the Canny and Hough transforms. These transforms do not have the ability to handle the noise reduction procedure, simultaneously. Moreover specific transforms such as Median filters, Gaussian blur procedure, or even FFT transform eliminate the noise while they dramatically reduce the details of the image (edges). In recent years considerable interest has been generated for wavelet transforms which is based on multiresolution analysis, time-frequency analysis, and pyramid algorithms [2].

In images, edge appears as the point where great changes in brightness are observed, separating different objects or different conditions. More specifically, it is of crucial importance in Remote Sensing to determine and detect the boundaries (lines) of significant objects

Humans can easily distinguish the limits of different objects in an image using different types of information like brightness, text, color as well as their knowledge of the world where the noise is isolated. However, this human understanding procedure is difficult to be automated.

Wavelet analysis may perform well in high-resolution images as it can manage better their detailed information and simultaneously reduce the noise. Edges of large objects are maintained in higher level scales whereas edges of smaller objects are maintained only in smaller levels of wavelet analysis. On the other hand, noise is usually introduced in the smaller levels. Thus, it is of high importance to establish a specific threshold policy which is related to the value that characterizes noise against information (edge) in the levels of wavelet analysis.

Edge detection techniques that are based on Wavelet analysis have been introduced since 1992. Mallat and Zhong [5], Zang and Bao [11], Sun et al. [10], and Shih and Tseng [9] introduced various methodologies of wavelet implementation where multiple scales should be employed to describe the variety of the edge structures along with the choice of the relevant wavelet mother and the threshold level.

It was pointed out that if an edge detector is to detect the zero-crossings as edge points, it must be symmetric with respect to the origin. On the other hand, if an edge detector is to detect the local extrema as edge points, it must be antisymmetric with respect to the origin [3].

Finally, as suggested by Marr and Hildreth [6], multiple scales should be employed to describe the variety of the edge structures and then these multiscale descriptions could be synthesized to form an edge map. Thus, wavelet analysis being a multiscale analysis can be used in edge detection.

11.4   Scale Multiplication

The scale multiplication can enhance image structures and suppress noise. An integrated edge map will be formed efficiently while avoiding the ill-posed edge synthesis process, unlike other multiscale edge detectors, where the edge maps are formed at several scales and then synthesized together. It was shown [11] that much improvement is obtained on the localization accuracy and the detection results are better than using either one of the two scales independently. Moreover, it will be improved more combining horizontal, vertical, or diagonal coefficients.

In order to reduce and smooth the existing noise an increase in the filter scale is required. An edge could disappear or could be dislocated if there is another edge curve at its neighborhood. It was also found that the scale multiplication will significantly reduce the interference of neighboring edges.

The peaks due to edges tend to propagate across scales, thus by directly multiplying the DWT at adjacent scales will enhance the edge structures whereas it will dilute the noise. With scale varying along dyadic sequence $2^j, j \in Z$, the support of wavelet base $\psi_j(x)$ will increase rapidly. This is also to say $W_j f(x)$ will become smoother rapidly along scales. If three or more adjacent scales were incorporated in the multiplication, edges would not be sharpened more but much edge dislocation would occur. So it is appropriate to analyze the multiplication using two scales [11].

In Fig. 11.2(a), a block signal g and its noisy version $f = g + \varepsilon$ are illustrated where ε is Gaussian white noise. Their DWT at the first three scales are given in Fig. 11.2(b) and (c). It is shown that at the finest scale the wavelet coefficients $W_1 f$ are almost dominated by noise. At the second and third scales, the noise diluted rapidly. It can also be seen that at the small scales the positions of the step edges are better localized. But some noise may be falsely considered as edges. At the large scales, the SNR is improved and edges can be detected more correctly but with the decrease of the accuracy of the edge location. In Fig. 11.2(d), the product $P_j^f, j = 1\text{--}3$, are illustrated. Apparently the step edges are more observable in P_j^f than in $W_j f$ [11].

In this scheme, the single threshold is preferred for the simplicity, as edges and noise can be better distinguished in the scale product and a properly chosen threshold could suppress the noise maxima effectively.

The edges are considered the local maxima in P_j^f. A significant edge at abscissa x_0 will occur on both the adjacent scales and the signs of $W_j f(x_0)$ and $W_{j+1} f(x_0)$ will be the same, so that $P_j^f(x_0)$ should be nonnegative. If $P_j^f(x)$ is less than zero, the point will be considered as noise and filtered out.

The scale multiplication will improve the detection performance (especially on the localization accuracy) and reduce the interference of neighboring edges [11] and thus it reduces the noise of the images.

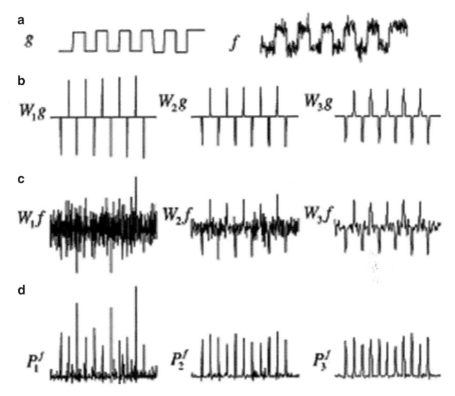

Fig. 11.2 (**a**) A signal g and its noisy version f, (**b**), (**c**) DWT of g and f, respectively, in the first three scales (**d**) the results of scale multiplication in f. (source: [11])

11.5 Evaluation Process

11.5.1 Introduction

In this paper an IKONOS pan-sharpened image of a suburban area in Agios Stefanos region (Athens, Greece) was used. Different data sets of this scene, containing man-made objects were selected and different band combinations were used.

Additionally an ALOS PRISM (panchromatic) image of Antiparos, Greece used in order to study the capability of wavelet transform in noise reduction field (not related to edge detection).

The scale multiplication scheme was chosen because it is based on wavelet analysis and gives better results in noisy images. Considering the non-desired detailed information of high-resolution satellite images as noise, which is reduced in higher levels of wavelet transform, the scale multiplication scheme would possibly perform well.

Fig. 11.3 Original Image
(IKONOS, SpaceImaging)

11.5.2   Scale Multiplication

The capability of the biorthogonal wavelet rbio3.1 was introduced, as it is asymmetric and resembles the wavelet proposed by Mallat [8].

Scale multiplication was primarily tested in artificial noisy images (containing straight lines of different directions and curved lines) as it is easier to evaluate the performance of the method and compare it to other ones in artificial images and then use it with the real, more complex, satellite images.

These tests showed that the method performs well in the multiplication of lower levels. In higher levels a significant dislocation of edges occurs. The levels 2–3 showed the best performance, presence of noise, but it was not able to detect edges which are very close to each other, due to the lower resolution on these levels. Levels 1–2 perform well when noise does not exist. Results from scale multiplication in noisy images were better than those of the Canny edge detector.

After these general considerations, a more detailed study was carried out using the IKONOS images from different regions (urban and suburban areas) were used to evaluate the "scale multiplication" performance in high-resolution satellite images. The scale multiplication scheme, gave good results in detecting manmade objects like buildings or road sides, without being disturbed from the detailed information of high-resolution satellite images. In Fig. 11.3 the image presented is the original images used, the 321 composite of the pansharpened image transformed in grayscale (Fig. 11.4). The levels 1,2 of the wavelet transform were used in scale multiplication, because the buildings and roads in this image are not big enough to use higher scales. In higher scales, a distortion of their shape is observed or in other cases they cannot be detected well. Thus small scales were preferred. It was found that edges of

Fig. 11.4 Detected edges
with rgb-image in the
background

Fig. 11.5 Canny edge
detector

manmade features were detected while in open areas; where small trees exist, trees edges were not detected. The results were slightly better than those of the Canny edge detector (Fig. 11.5), but they could be further improved, by better choice of the mother wavelet (Fig. 11.6).

Four levels of the transform were used as it was observed that using only the coefficients of lower scales, the results were not really good as the features of

Fig. 11.6 Edges resulted
from scale multiplication

the reconstructed image usually are broken, because some of the information is contained in the next level. However, in order to detect big features or objects the coefficients of lower levels are required.

Then the reconstructed image was thresholded to retain only the most significant values. The produced image was a binary image that represents the location of high coefficients in the contourlet domain.

Finally, a further cleaning of the image was performed to retain only the road network, by using other morphological operators like fill, thin, and shrink.

In the reconstructed image (without the participation of the approximation coefficients), the higher values represent the locations where the radiometry increases whereas the lower (negative) values represent the location where the radiometry decreases. Both, high positive and negative values constitute edge points. To maintain the negative values the threshold should take into consideration the absolute value. Moreover, two edge images can be created, one for the positive values and one for the negative, by using different thresholds. The final edge map would result from the synthesis of the other two maps (Figs. 11.7–11.12).

11.6   The ALOS PRISM Image and the Noise Reduction

This case is very challenging, as ALOS/PRISM images do have radiometric quality problems, leading to image noise which could be reduced without influence to the image resolution [7]. This noise can influence in a negative way the geo-reference procedure (measurements of GCPs and Independent Check Points) the image matching for DSM generation and the image interpretation procedures general.

Fig. 11.7 IKONOS
Pansharpened, 321 composite
(SpaceImaging)

Fig. 11.8 Reconstructed
image after coefficient
reduction

PRISM data are separated into odd and even detectors and transmitted from satellite by different transfer channel. Odd and even detectors are compressed independently in JPEG compression. The JPEG compresses an image block by block; a block is 8 pixels by 8 lines. Therefore a block of 16 pixels by 8 lines on uncorrected image is consists of 2 JPEG blocks: a block of odd pixels and a block of even pixels.

Fig. 11.9 Edge map

Fig. 11.10 Edge map after
morphological operations

The noise which appears on ALOS PRISM CCDs is referred as a stripe noise
focus mainly on solving the issue of the brightness difference between odd and even
pixels. This noise could be caused due to independent transfer of data of odd and
even detectors.

This method is based on wavelet analysis with four levels as introduced in the
previous paragraphs, it is possible to match and control the spatial differences

Fig. 11.11 IKONOS, band 4
(SpaceImaging)

Fig. 11.12 Edge map after
morphological operations

between the odd and even detectors, without loss of the image detail and information. In Figs. 11.13 and 11.14 crops of the ALOS nadir images are introduced
while in Figs. 11.15 and 11.16 are the images where the specific filter is applied.
The results are very promising as the noise is dramatically reduced, while on the
other hand the textures and the image detail are still there.

Fig. 11.13 Part of the ALOS
PRISM nadir image. The
noise dominates the image

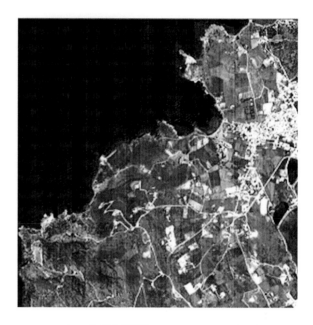

Fig. 11.14 Part of the ALOS
PRISM nadir image. The
noise dominates the image

11.7 Conclusions

The two main advantages of wavelet analysis are that it is multiscale analysis and
at the same time local analysis. An image can be decomposed into components
containing information of different frequencies or in other words information of

Fig. 11.15 The image of
Fig. 11.1 after the noise
reduction

Fig. 11.16 The image of
Fig. 11.14 after the noise
reduction

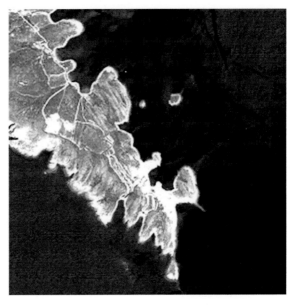

different scales without losing its location. Additionally, the method is really flexible
as it provides the means to use different mother wavelets and also to construct a new
one according to the application. These characteristics constitute the use of wavelet
analysis in remote sensing and image processing an efficient method for denoising,
image fusion, edge detection, etc.

More precisely wavelet analysis can be used in edge detection and noise reduction of high-resolution satellite images where the notion of scale is really important. It was shown that the scale multiplication of wavelet analysis can be used in detecting the boundaries of buildings or road sides, avoiding the disturbance of detailed information such as small trees, cars, and other small objects. Also, it is shown that the wavelets can be an efficient tool for reducing even periodical noise, as in the case of ALOS PRISM. The performance of the method can be improved by using more efficient wavelets for specific applications, or different schemes based on wavelet analysis like those using statistical properties of wavelet coefficients, to determine edges.

References

1. Castelman, R.: Digital Image Processing, In: Prentice Hall, Englewood Cliffs, NJ, 2nd edn. p. 667 (1995)
2. Gonzales, R., Woods, R., Eddins, S.: Digital Image Processing using Matlab, In: Pearson Education, New Jersey, p. 620 (2004)
3. Hsieh, J.W., Ko, M.T., Liao, H.Y.M., Fan, K.C.: A new wavelet-based edge detector via constrained optimization. Image Vis. Comput. **15**, 511–527 (1997)
4. Mallat, S.: A Wavelet Tour of Signal Processing, In: Academic Press, San Diego, 2nd edn. pp. 637 (1998)
5. Mallat, S., Zhong, S.: Characterization of signals from multiscale edges. IEEE Trans. Pattern Anal. Mach. Intell. **14**, 710–732 (1992)
6. Marr, D., Hildreth, E.: Theory of edge detection. Proc. Roy. Soc. Lond. B Biol. Sci. **207**(1167), 187–217 (1980)
7. Nikolakopoulos, K., Tsombos, P., Michalis, P.: Spatial resolution enhancement of ALOS data for geological mapping, ALOS Joint PI Symposium 2008, Rhodos-Greece, 3–7 November 2008
8. Noutsou, V., Argialas, D., Michalis, P.: Edge detection of man made objects using wavelet domain in high resolution satellite images. ASPRS (American Society of Photogrammetry and Remote Sensing) Annual Conference (2007)
9. Shih, M., Tsehng, D.: A wavelet-based multiresolution edge detection and tracking. Image Vis. Comput. **23**, 441–451 (2005)
10. Sun, J., Gu, D., Chen, Y., Zhang, S.: A multiscale edge detection algorithm based on wavelet domain vector hidden Markov tree model. Pattern Recogn. **37**, 1315–1324 (2004)
11. Zang, L., Bao, P.: Edge detection by scale multiplication in wavelet domain. Pattern Recogn. Lett. **23**, 1774–1784 (2002)

Chapter 12
Optimal Orbital Coverage of Theater Operations and Targets

Vasileios Oikonomou

Abstract The use of satellites as a tactical asset to support theater operations is a desired capability for future space operations. Unlike traditional satellite systems designed to provide coverage over the entire globe or large regions, tactical satellites would provide coverage over a small region which can be modeled as a single ground point defined by a latitude and longitude. In order to provide sufficient utility as a theater asset, a satellite should be placed in an orbit that provides a maximum amount of coverage of the target ground point. This study examined the optimization of orbit parameters to maximize the number of passes made over a target. An optimization algorithm was developed to maximize the number of passes made while also minimizing the distance from the satellite to the target. Single satellite coverage properties as well as two and three satellite constellations were analyzed.

Keywords Satellite • Coverage of a theater target • Orbit • Latitude • Longitude • Analytical approach • The number of daylight passes • The slant range to the target • Coverage geometry

12.1 Tactical Satellites and Responsive Space Operations

Space assets have become an important and often critical part of military operations. Satellites are employed in a variety of missions including surveillance, communication, and navigation. Currently, satellite systems are managed as national assets but there is a strong interest in developing satellites that would be managed as a tactical

V. Oikonomou
Alikarnassou 102A, GR 85300, Kos Greece
e-mail: billoikonomou@hotmail.com

N.J. Daras (ed.), *Applications of Mathematics and Informatics in Military Science*,
Springer Optimization and Its Applications 71, DOI 10.1007/978-1-4614-4109-0_12,
© Springer Science+Business Media New York 2012

asset. Using space assets as a tactical tool is part of a strategy termed responsive space operations. The concept of responsive space operations focuses on the ability to launch space assets in response to an emerging threat or identified need. A tactical satellite, launched in support of a planned or ongoing theater operation, would be under the control of the theater commander and make space capabilities a tactical asset. Tactical satellites could fill various missions such as providing additional surveillance of the theater or augmenting communications systems. As stated by General Cartwright, tactical satellites must "demonstrate that operationally relevant, rapidly deployable spacecraft can support military operations anywhere on Earth." In order to be rapidly deployable, tactical satellites will most likely be launched into low Earth orbits. This will allow the satellites to begin on-orbit operations in a minimal amount of time. In order to be operationally relevant, a satellite will need to provide a sufficient amount of utility for its mission area. For surveillance, an important measure of a satellite's utility is the coverage it provides of the target area. In order to provide the most utility, a satellite should be placed in an orbit that maximizes the coverage of the theater target. Designing orbits that maximize the coverage of a specified target is an important research area for the concept of tactical satellites.

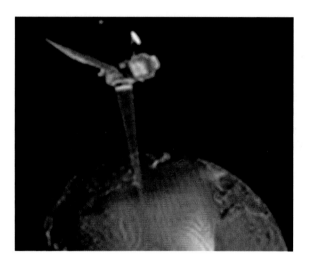

12.2   Research Objectives

The primary objectives of this research were to find methods of optimizing the coverage of a theater target by a satellite and determining the optimum orbit parameters for the satellite. A theater target was specified by a latitude and longitude. Since the latitude of the target will vary, it was also desired that the

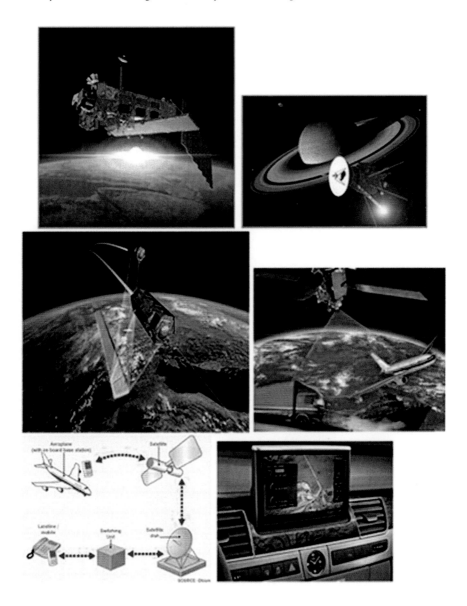

effects of latitude location on coverage optimization be observed. The coverage of the target was measured as the number of daylight passes made over the target. Since tactical satellites might be employed as a single satellite or a small constellation, constellation design was also addressed.

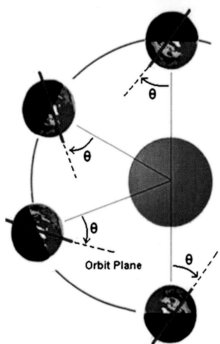

12.3   Assumptions and Limitations

The use of satellites as a tactical asset imposes a practical limitation on the orbit altitude. The research included only analysis of low-Earth orbits. Since satellites designed for tactical use will need to be launched, placed in orbit, and operating in a timely manner, low Earth orbits are the most practical and suitable choice. Both circular and elliptical orbits were compared for performance in test cases, but only circular orbits were optimized and used for constellation design. For the purposes of this research the primary mission of the satellite was assumed to be collection of visible imagery of a theater target. Surveillance of a target may take various forms, but visible imagery is a common and highly valuable resource for theater operations. Visible imaging requires that the target be illuminated by the sun, which limits a satellite to daytime operations. Since this is an important constraint on the imaging opportunities, only passes made over the target during daylight were included in the coverage analysis for a satellite. The research focuses on optimizing coverage over a limited time period. Since the satellites will be providing tactical support for theater operations, it was assumed the mission duration would be short. A nominal period of 30 days was chosen as the time period. Since a satellite's longitude of ascending node regresses over time due to orbital perturbations, the value selected for the node represents the initial value at the beginning of the 30 day period. The value is selected to allow the node to drift through its optimal value during the time period [1].

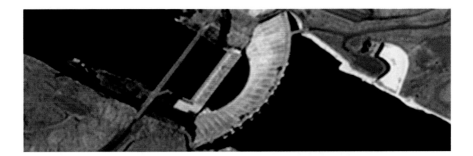

12.4   Methodology

In order to analyze the coverage properties of a satellite, an analytical approach was used to characterize the problem. A satellite's coverage of a target will be largely affected by its field of view which in turn is determined by its orbital altitude. The relationship between a satellite's field of view, orbital altitude, inclination, and the latitude of the target is examined analytically. The analysis is limited to circular orbits and several simplifying assumptions are made. In order to analyze the problem more accurately a numerical method was used. In order to assess the coverage performance of a satellite, a computer program was designed to simulate the scenario. The simulation included an orbit propagator which used a numerical method to simulate the dynamics of a satellite and measure its position and velocity over the specified time period. The orbit propagation included the effects of the J2 perturbation caused by the oblateness of the Earth. The simulation propagated the position of the target site in inertial space by simulating the rotation of the Earth and propagated the position vector from the sun to the Earth. Using the simulated scenario, the program determined how many passes the satellite made with the target visible to the satellite and illuminated by the sun. The program measured the number of daylight passes the satellite made over the target, the length of each pass, the slant range from the satellite to the target during each pass, and the time at which passes occurred. In order to optimize the performance of a satellite, the number of daylight passes made was chosen as the primary figure of merit. For a given orbital altitude and inclination, the longitude of ascending node was optimized to the value that provided the highest number of daylight passes. For a given altitude, the inclination which provides the highest number of daylight passes was determined and the associated trends examined. Constraints including a minimum elevation angle or a maximum slant range were also examined. The test cases at varying altitudes revealed a tradeoff between the number of daylight passes and the average slant range to the target. Since the slant range will affect the resolution of visible imagery it was added as another figure of merit. The number of daylight passes and average slant range were selected as the performance criteria for orbit optimization. Maximizing the number of passes was chosen as one objective in order to provide the most opportunities for the satellite to capture imagery of the target. Minimizing the slant range to the target was chosen as the other objective in order to allow the highest resolution for the imagery. A multi-objective optimization algorithm using a weighted cost function was designed to select optimal orbits to meet the coverage objectives. Using one of the optimized orbits, constellation design was examined. Two satellite constellations were designed by varying the mean anomaly of the second satellite and by varying the longitude of ascending node of the second satellite. Three satellite constellations were designed using the same techniques. A constellation was also designed for extended operations [3].

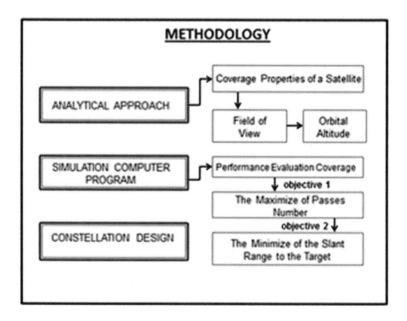

12.5   Analytical Analysis

The first approach used to examine the problem was an analytical analysis of the coverage properties of satellites. Satellites in circular orbits lend themselves well to analytical analysis because the coverage properties do not vary as the satellite travels around the Earth. Although an analytical approach provides insight, it also has limitations. Analytical approximations do not take into account orbital perturbations or other factors that will play an important role in the coverage properties of a real satellite.

12.5.1   Coverage Geometry

The coverage geometry for a satellite is depicted in figure.

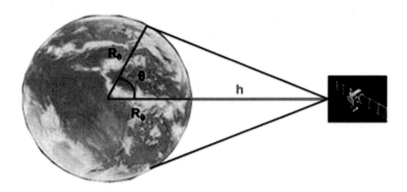

A satellite's field of view on the Earth's surface will be a circle and the size of the field of view will be determined by the satellite's altitude. For a satellite in a circular orbit, the field of view will remain a constant size while a satellite in an elliptical orbit will have a field of view whose size varies as the satellite's altitude changes. The Earth central angle (θ) can be used to describe the size of half the satellite's field of view. For a satellite in a circular orbit and using a simplified spherical Earth model, the Earth central angle can be determined from the equation. The equation can also be used to determine the Earth central angle for a point on an elliptical orbit with a particular altitude, h.

$$\cos \theta = \frac{R_\theta}{R_\theta + h}.$$

The field of view shown in figure is limited only by the satellite's altitude which determines where the local horizon is and thus the farthest point that is within the satellite's field of view. Many satellites have additional constraints on their field of view. There may be a minimum elevation angle required for the satellite to operate effectively due to obstructions in its line of sight. The satellite might have a maximum operating slant range for its onboard instruments.

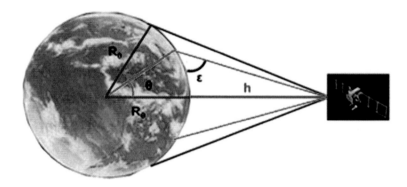

The figure depicts an elevation angle (ε) constraint which creates an effective field of view limited further than the horizon. If the field of view is constrained by an elevation angle requirement, the Earth central angle can be determined using the equation:

$$\cos(\theta + \varepsilon) = \frac{R_\theta \cos \varepsilon}{R_\theta + h}.$$

The figure depicts the coverage for a satellite with a maximum slant range (ρ).

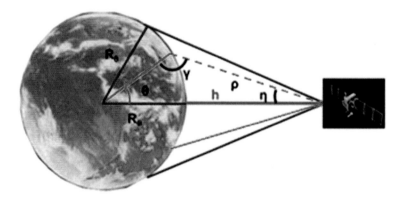

If the field of view is constrained by a maximum slant range, the Earth central can be determined from the equations:

$$\Theta = 180 - (\eta + \gamma)$$

$$\sin \gamma = \frac{(R_\theta + h) \sin \eta}{R_\theta}$$

$$\cos \eta = \frac{\rho^2 + (R_\theta + h)^2 - R_\theta^2}{2\rho(R_\theta + h)}.$$

12.6   Target Coverage

A satellite in low Earth orbit which seeks to cover a certain target latitude and longitude and which has an inclination near the target latitude $((\theta - L)$ to $(\theta + L))$ will follow a pattern in which it makes a series of successive passes during which it has coverage of the target followed by a number of passes during which it has no coverage of the target. This pattern will continue to repeat itself. The range of longitude that the satellite covers will affect the number of successive passes that have coverage of the target. For a satellite in a given orbit with period, P, the ground track of the satellite will appear to shift westward in longitude with every orbit pass because the orbit is fixed in inertial space while the Earth is rotating eastward. The shift (s) can be measured using the equation:

$$S = P \cdot \omega.$$

For an orbit with a longitude range 2ϕ, if l_1 represents the beginning of the longitude range (eastern most longitude) and $l_2 = l_1 + 2\phi$ represents the end (western most longitude) of the longitude range, then after one pass

$$l_1 = l_1 - s$$
$$l_2 = l_2 - s.$$

If the target longitude l_t was exactly at l_1 for a given pass,

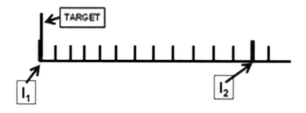

then on the subsequent pass the target would be located at $l_1 + s$

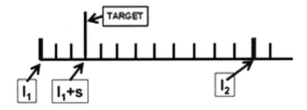

And on the following pass at $l_1 + 2s$

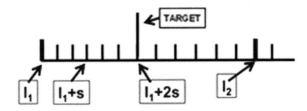

And so on until $l_1 + xs > l_2$.

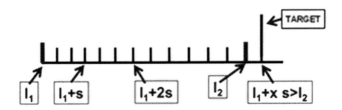

At this point the target longitude is no longer within the longitude range and cannot be seen by the satellite. For the case described above the number of successive passes is given by

$$\#\text{Passes} = \left(\frac{2\phi}{s} + 1\right).$$

However this is only for the case where the target longitude was initially at l_1 If the target longitude is initially at $l_1 + \sigma$ (where $\sigma < s$).

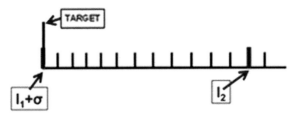

then on the following pass the target longitude would be located at $l_1 + s + \sigma$.

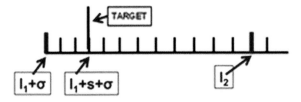

And on the next pass at $l_1 + 2s + \sigma$.

And so on until it reaches the point where $l_1 + x \cdot s + \sigma > l_2$.

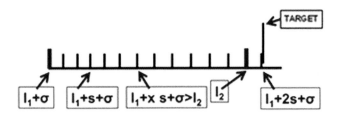

The number of passes will be given by [4]

$$\#\text{Passes} = \left(\frac{2\phi}{s} - \frac{\sigma}{s} + 1 \right).$$

12.7 Computer Simulation

The analytical analysis applies to a simplified case and has various limitations. It does not take into account whether a target site is in daylight which is critical for visible imaging systems. It also does not include the effects of orbital perturbations and allows only for the examination of circular orbits. For further analysis a computer simulation was used. The simulation propagated a satellite over a thirty day time period and measured its coverage of a selected target site. The coverage was measured as the number of daylight passes made by the satellite while within view of the target site. The J2 orbital perturbation was simulated to include its orbital effects. Satellites in both circular and elliptical orbits were simulated as well as constellations of satellites.

For the computer simulation were used the following quantities.

(a) Sun position vector
    In order to determine if the target site was in daylight, the position vector from the Earth to the sun was needed. The position vector from the Earth to the sun is an Earth-centered inertial frame.

(b) Site Position vector
    In order to determine if the target site was within view of the satellite, the target site's position vector was required. The target site's latitude, longitude, and altitude were used to determine its position vector in an Earth-centered Earth-fixed coordinated frame using the semimajor axis and eccentricity of the Earth.

(c) Site Illumination
    To determine whether the target site was illuminated by the sun, the angle between the sun's position vector and the site's position vector was calculated.

(d) Slant Range
    The slant range is the distance from the satellite to the target.

(e) Site Visibility
    The position vector from the target site to the satellite and the elevation angle from the site to the satellite were used to determine whether the target site was within view of the satellite. If the elevation angle was greater than the minimum required elevation angle (or zero if no minimum elevation angle had been designated) then the site was considered to be visible to the satellite [5].

12.8   Optimization Algorithm

The number of daylight passes made over the target and the average slant range from the satellite to the target are two important coverage properties. The maximum number of daylight passes will provide the maximum number of opportunities for imaging of the target. The average slant range will affect the resolution of the imagery and the minimum slant range distance will provide the highest resolution imagery. An orbit which provides a high number of daylight passes usually also has a high average slant range. In order to balance the trade-offs between the number of passes and slant range, an optimization algorithm was developed and implemented as a computer program.

Table 12.1 Optimization algorithm input parameters

Target parameters	Search span	Time parameters	Weighting parameters
Target latitude	Maximum orbit altitude	Time span	Maximum passes weight
Target longitude	Minimum orbit altitude	Start date	Minimum slant range weight
Target altitude			
Minimum elevation angle			

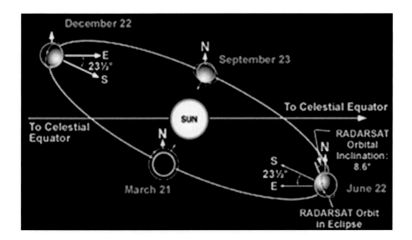

The algorithm takes in a series of inputs as shown in Table 12.1.

- The maximum and minimum altitudes define the span of orbit altitude that the algorithm will search over to select an optimum orbit.
- The time span specifies the length of time (in days) that the simulation will optimize over for an orbit and the start date specifies the date on which the time period will start. The algorithm only examines circular orbits and is intended only for low Earth orbits. It is also assumed that the target latitude is greater than the Earth central angle of each altitude in the search span.
- The weighting parameters are used to indicate the importance that should be given to the number of daylight passes made and to the average slant range [12].

12.8.1 Solution Space

The solution space for the optimization problem is found by determining the approximate maximum number of daylight passes and corresponding average slant range as well as the approximate minimum average slant range and corresponding number of daylight passes.

In order to find the minimum average slant range, the minimum orbit altitude as specified by the input parameter is used.

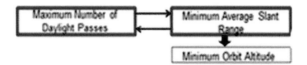

The program simulates a satellite at the minimum orbit altitude and an inclination equal to the target latitude. The longitude of ascending node is optimized to the value that provides the maximum number of daylight passes.

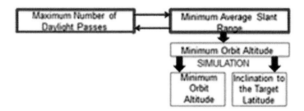

The average slant range is calculated as well as the number of daylight passes made.

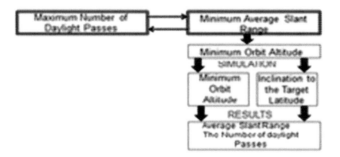

The program then increments the inclination value above the target latitude and again simulates the satellite and measures the average slant range and number of daylight passes made. The program continues to increment the inclination until a local minimum value for average slant range is determined for the minimum orbit altitude. This local minimum is considered the minimum value for average slant range (range min) for the solution space and the corresponding number of daylight passes is considered the minimum number of daylight passes (pass min) for the solution space.

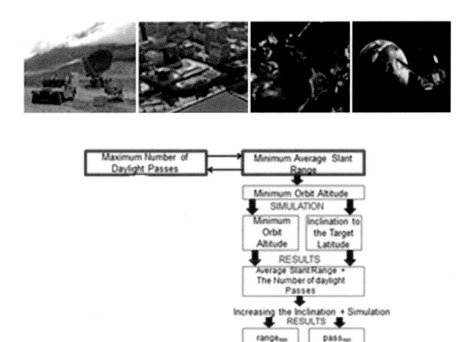

The maximum number of daylight passes is determined by using the maximum orbit altitude.

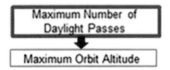

The longitude of the ascending node is optimized to provide the highest number of daylight passes. The number of daylight passes made is measured as well as the average slant range. The program then increments the inclination below the approximated value and simulates a satellite at the maximum orbit altitude and the new inclination value. This process is repeated until a local maximum value is found for the number of daylight passes. This local maximum is considered the maximum number of passes (passmax) for the solution space and the corresponding average slant range is considered the maximum slant range (rangemax) for the solution space.

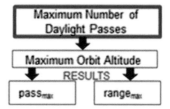

The maximum and minimum bounds of the solution space are used to determine the span of passes and span of slant range as shown in the equations:

$$\text{span}_{\text{pass}} = \text{pass}_{\text{max}} - \text{pass}_{\text{min}} \quad \text{span}_{\text{range}} = \text{range}_{\text{max}} - \text{range}_{\text{min}}.$$

The spans of average slant range and number of passes are then used to find a scaling parameter so that changes in range can be compared with changes in the number of passes. The scaling parameter is used to normalize the slant range and number of passes. Where δ represents the accuracy of the number of passes as determined by a simulation and is set at a default value of 5. The δ parameter is used to set a significance level for the number of passes. An increase of a single pass may not truly represent a better coverage property but could be the result of where the simulation stopped so a minimum of five passes is used to ensure that the difference in coverage is significant.

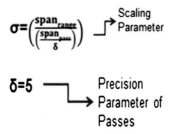

12.8.2 Optimization Solution Method

The optimization algorithm has two objectives, to maximize the number of daylight passes and to minimize the average slant range. To satisfy both objectives, a weighted cost function is used to find an optimal solution. A weighting parameter for the number of passes and a weighting parameter for the average slant range are used to determine the importance of each objective (Table 12.2). By changing the weighting parameters, different solutions can be found. The equation shows the cost function for the algorithm.

Table 12.2 Optimization algorithm output

Orbit parameters	Orbit coverage properties
Altitude	Number of daylight passes
Inclination	Total coverage time
Longitude of ascending node	Average pass length
	Average slant range
	Maximum slant range

$$C = \sum_{i=1}^{30} \lambda_1 \frac{x_1}{\delta} - \lambda_2 \frac{x_2}{\sigma},$$

where λ_1 = Weighting parameter for number of passes, λ_2 = Weighting parameter for average slant range, x_1 = Number of Daylight Passes, and x_2 = Average Slant Range [2].

12.9   Results

12.9.1   Introduction

The test cases presented reveal trends associated with orbit coverage properties. A key result of the test cases is the trade-off between the number of satellite passes made and the slant range from the satellite to the target. Since both of these properties are important for high resolution imagery, an optimization algorithm was used which takes into account and weighs the objectives of maximizing the number of passes and minimizing the slant range. Using one of the optimized orbits, an examination of constellation properties was performed to design configurations for constellations of two and three satellites.

12.9.2 Optimum Inclination for Maximum Number of Daylight Passes

Various test cases were examined to determine the inclination that maximizes the number of daylight passes made. For visible imaging satellites, an orbit that provides the most opportunities to capture imagery of the target is desired. The average slant range to the target was also measured since it is another important consideration for target coverage. The test cases include varied orbit altitude, target latitude, slant range constraints, and elevation angle constraints.

12.9.2.1 Analytical Predictions

The analytical analysis showed a direct correlation between the inclination of a satellite's orbit and the swath of longitude at a given latitude value that would be viewed each time the satellite completed an orbit around the Earth. An inclination at the value of the satellite's Earth central angle plus the target latitude provided the largest swath of longitude. In addition a larger longitude swath was shown to correspond to an increased number of passes over a target latitude and longitude. These results imply that the maximum number of passes will be made at an inclination equal to the Earth central angle plus the value of the target latitude. To test this prediction, various orbit cases were used in the computer simulation.

12.9.2.2 350 km Altitude Circular Orbit, Target Latitude 33°

In order to determine the optimum inclination for a satellite in a 350 km circular orbit, a range of inclinations was tested to see where the maximum number of daylight passes occurs. The tests were run using a 30 day time period and the total number of daylight passes measured. For each pass counted, the target site was illuminated by the sun and therefore in daylight and the target site was visible to the satellite. The target site was considered visible if the elevation angle was greater than or equal to 0°. For each inclination tested, the longitude of the ascending node was optimized to yield the highest number of daylight passes. Figure shows the results for a target at a latitude of 33°. As expected, the number of daylight passes varies depending on the inclination. The number of passes increases as inclination is increased until it reaches its maximum at 51°. This optimum inclination provided 186 passes during the 30 day time period. The number of passes then drops off steeply as the inclination is increased above 51°. The optimum inclination occurs at 18° above the target latitude and the Earth-central angle for an orbital altitude of 350 km is 18.56°. Thus the analytical prediction that the optimum inclination occurs at the target latitude is consistent with the results.

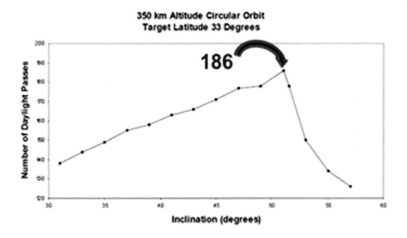

12.9.2.3 800 km Altitude Circular Orbit, Target Latitude 33°

The second case tested was an 800 km altitude circular orbit. The target latitude was kept at 33° and a range of inclinations was tested to determine the inclination at which the maximum number of daylight passes occurs. The minimum elevation angle used was 0° and the time span 30 days. For each inclination tested the longitude of the ascending node was optimized to the value that yielded the highest number of daylight passes. Figure shows the number of daylight passes for the values of inclination tested.

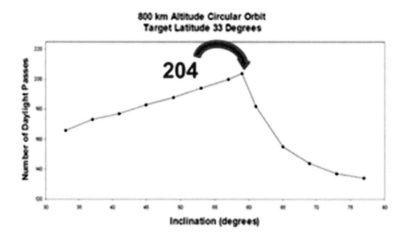

The maximum number of daylight passes made over the 30 day period is 204 passes which occurs at an inclination of 59°. The trends shown are consistent with the 350 km altitude orbit. The number of passes increases as the inclination is increased above the target latitude until it reaches a maximum at 59°. As the inclination is increased above 59° the number of passes decreases quickly. The Earth-central angle for an 800 km altitude orbit is 27.31° and the optimum inclination is 26° above the target latitude. In comparison with the 350 km altitude orbit case, an increased number of daylight passes are made by the 800 km altitude case. This is an expected result since increasing the altitude of an orbit increases the satellite's field of view on the surface of the Earth. The total coverage time and average pass length have increased in comparison to the 350 km case, but the average and maximum slant range have also increased. The average slant range versus inclination for the 800 km case is shown in the next figure.

The average slant range is at its maximum at the optimum inclination for maximum number of daylight passes. The trend is consistent with the trend seen for the 350 km altitude orbit case. As the orbital altitude is increased, the number of daylight passes made over the 30 day period also increases.

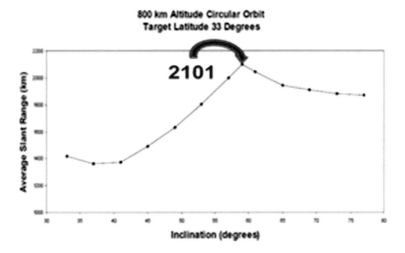

350 km Altitude Circular Orbit, Target Latitude 0°

12.9.2.4   350 km Altitude Circular Orbit, Target Latitude 0°

The case of a target site on the equator was tested over a range of inclinations. Each test was run for a 30 day time span and the total number of daylight passes recorded. For each inclination tested the longitude of the ascending node was optimized to provide the highest number of daylight passes. The minimum elevation angle used from the target site to the satellite was 0°. Figure shows the results for a 350 km altitude circular orbit and a target latitude of 0°.

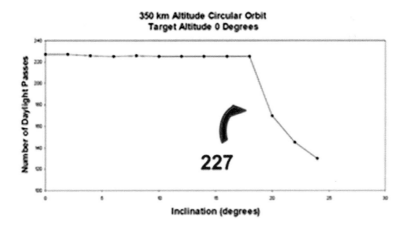

As the inclination is increased above the target latitude there is no significant change in the number of passes until the inclination reaches 18°, at which point the number of passes drops off steeply as the inclination is increased. The trend suggests there is no improvement in increasing inclination above the target inclination. At an inclination of 0°, the satellite makes the maximum number of daylight passes which is 227 passes. For a satellite in an equatorial 350 km altitude orbit, the latitude of the target is always within view of the satellite so no additional benefit can be gained by increasing the inclination. The Earth-central angle is 18.56° for a 350 km altitude circular orbit. As long as the inclination remains at or below 18°, the latitude of the target site should always be within view of the satellite. Next figure shows the average slant range for each of the inclinations tested.

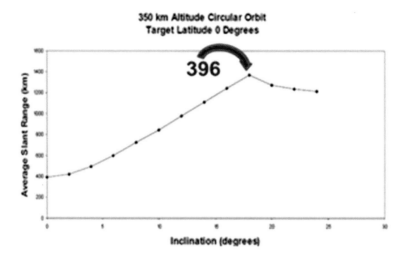

As the inclination is increased the average slant range also increases until it reaches its maximum at 18°. Although there is little change in the number of passes as long as the inclination remains below 18°, an inclination of zero offers the additional benefit of having the smallest average slant range to the target site. Although there may not be a significant decrease in the number of daylight passes at inclinations greater than zero, there is a significant increase in the average slant range from the satellite to the target.

12.9.2.5 800 km Altitude Circular Orbit, Target Latitude 0°

An 800 km altitude circular orbit case was also tested with the target site placed on the equator. A range of inclination values was tested and the number of daylight passes measured. Each inclination was tested over a 30 day time span and a minimum elevation angle of 0° was used. For each inclination tested the longitude of the ascending node was optimized to provide the highest number of daylight passes. The results for the number of daylight passes are shown in figure.

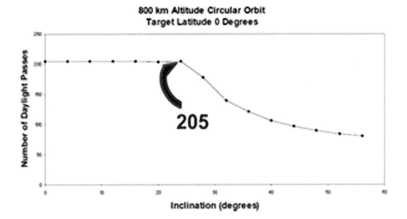

800 km Altitude Circular Orbit
Target Latitude 0 Degrees

The maximum number of daylight passes made over the 30 days is 205 passes. There is no significant variation in the number of passes made as the inclination is increased from zero to 24°. As the inclination is increased above 24° the number of daylight passes steadily decreases. The trend shown is consistent with the 350 km orbit case. As long as the satellite's inclination is at or below 24°, the target's latitude band will always be within view and there will be little variation in the number of passes. Above an inclination of 24° the number of passes will decrease as the inclination is increased. Next figure shows the average slant range versus inclination for this case. The trend is again consistent with the 350 km orbit case. As the inclination is increased, the average slant range increases until it reaches a maximum at 28°. The average slant range is minimized at an inclination of 0°.

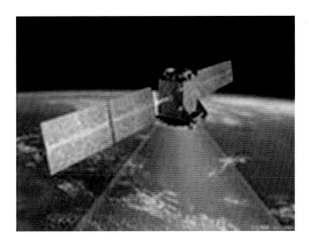

Table 12.3 Comparison
of Coverage Properties
Equatorial Orbits at 350 km
and 800 km

Orbit altitude (km)	350	800
Number of daylight passes	227	205
Total coverage time (h)	36.6	54.4
Average pass length (min)	9.7	15.9
Average slant range to target (km)	396	830
Maximum slant range to target (km)	2,073	2,945

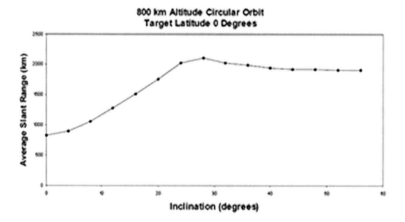

In comparison with the 350 km altitude orbit case, a fewer number of daylight passes are made by the satellite in an 800 km altitude orbit. Table 12.3 summarizes the coverage properties for a satellite in an equatorial orbit at 350 km altitude and at 800 km altitude.

The 800 km altitude orbit provides fewer total passes but does provide a larger total coverage time due to longer duration passes. The 350 km altitude orbit provides a greater number of passes and a smaller slant range to the target site.

12.9.2.6   350 km Altitude Circular Orbit, Target Latitude 10°

Another case tested was a circular orbit at 350 km altitude and a target site at 10° latitude. A range of inclinations was tested to determine the inclination that provides the maximum number of daylight passes. The tests were run using a 30 day time period and for each inclination tested the longitude of the ascending node was optimized to provide the highest number of daylight passes. The minimum elevation angle used was 0°. Figure shows the results for the number of daylight passes made during the 30 day period.

The maximum number of daylight passes made over the 30 day period is 236 passes. At an inclination of 0° the number of daylight passes made is the maximum amount. As the inclination increases above 8°, the number of daylight passes begins to steadily decrease. The average slant range is shown in the next figure. The slant range is minimized at an inclination of 14°.

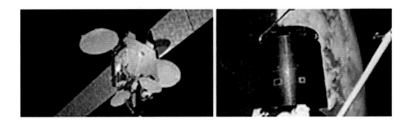

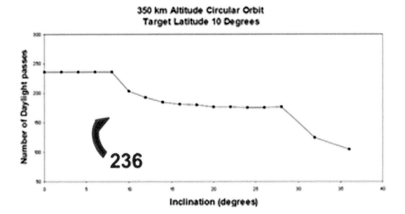

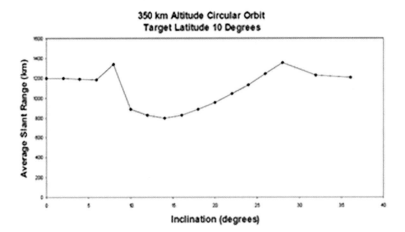

12.9.2.7   800 km Altitude Circular Orbit, Target Latitude 10°

An 800 km altitude orbit was tested with the target placed at a latitude of 10°. The tests were run for a time period of 30 days and a minimum elevation angle of 0° was used. The longitude of the ascending node for each orbit was optimized to provide the maximum number of daylight passes. For each inclination tested the number of daylight passes made over the target was measured. Figure shows the results for the range of inclinations tested. The maximum number of daylight passes made over the 30 day period is 214 passes. At an inclination of zero the maximum number of passes is made. The number of passes made remains the same until the inclination is increased above 16°, at which point the number of passes decreases and continues to decrease as the inclination is increased.

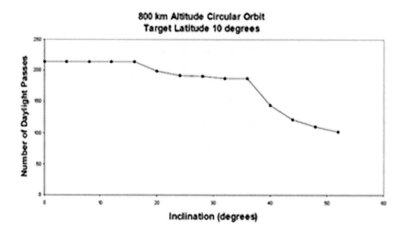

Next figure shows the average slant range for each inclination tested. The slant range is minimized at an inclination of 12° [6].

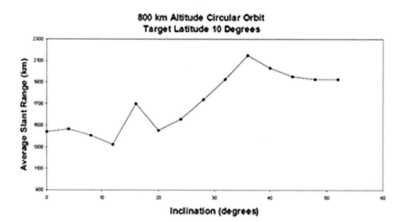

12.9.3   Optimum Inclination for Maximum Number of Daylight Passes with Constrained Slant Range

The slant range, measured as the distance from a satellite to a target along its line of sight, may be an important factor for satellite operations. The surveillance tools onboard the satellite may have a maximum distance at which they can effectively operate. When this is the case, any passes made over the target will only be useful if the slant range to the target is less than the maximum distance required by the equipment. In order to examine the impact of a maximum slant range on inclination optimization, several cases were run with a constraint placed on the slant range. Only satellite passes made with a slant range less than the constraint were counted during the simulations.

12.9.3.1   350 km Altitude Circular Orbit, Target at Latitude 33°, MaximumSlant Range 800 km

The first case tested was a 350 km altitude circular orbit and a target at a latitude of 33°. The slant range constraint chosen was a maximum slant range of 800 km. A range of inclinations was tested to see where the maximum number of daylight passes occurs. For each inclination tested the longitude of ascending node was optimized to provide the maximum number of daylight passes. Each test was run for a time period of thirty days. Figure shows the number of daylight passes made over the thirty day period.

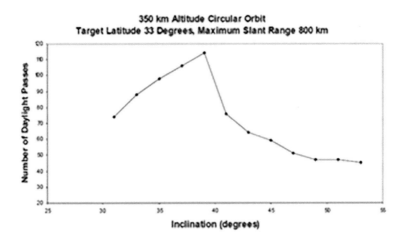

As the inclination is increased above the latitude of the target, the number of daylight passes increases until it reaches a maximum at 39°. The maximum number of passes made over the 30 days is 114 passes at an inclination of 39°.

As the inclination is increased above 39°, the number of passes made decreases. The average slant range is 594 km which is well below the constraint of 800 km. In comparison with the unconstrained case, the number of passes made is less. At an inclination of 39°, the unconstrained case of a 350 km altitude orbit would have yielded 158 passes. With the slant range constraint, 114 passes are made in which the slant range requirement is met and 44 passes are made at a distance that exceeds 800 km. The maximum number of passes occurs at an inclination of 39° in comparison with 51° for the unconstrained case. For a satellite at an altitude of 350 km and using a maximum slant range of 800 km, the effective Earth central angle is approximately 6.3° as depicted in figure and in this case the optimum inclination occurs at 6° above the target latitude; again the results are consistent with the analytical analysis [10].

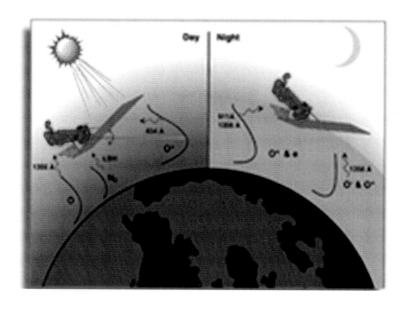

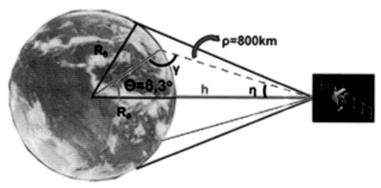

12.9.4 Optimum Inclination for Maximum Number of Daylight Passes with Constrained Elevation Angle

The elevation angle, measured from the local horizon of the target to the line of sight vector to the satellite, may have a minimum value that is greater than zero. All satellites are limited by the horizon but often a higher elevation angle is also required. For visible imaging satellites, an elevation angle of 90° is ideal because the image will be taken directly overhead. As the elevation angle decreases, the images will be more difficult to interpret and less useful. At very small elevation angles, objects may obstruct the line of sight of the satellite to the target and prevent it from operating. In order to examine the impact of a minimum elevation requirement, several cases were run with a constraint on the minimum elevation angle required to view the target. If the elevation angle was smaller than the constraint value, the target was not considered visible to the satellite.

12.9.4.1 350 km Altitude Circular Orbit, Target at Latitude 33°, Minimum Elevation Angle 10°

The first case examined was a 350 km altitude circular orbit with the target placed at a latitude of 33°. A minimum elevation angle of 10° was used. In order to determine the inclination at which the maximum number of daylight passes occurs, a range of inclinations was tested. A time period of 30 days was used and the longitude of the ascending node was optimized to provide the maximum number of daylight passes for each inclination tested. Figure shows the number of daylight passes made at each inclination tested. The maximum number of daylight passes made over the 30 day period is 144 and occurs at an inclination of 43°. As the inclination is increased above the target latitude, the number of passes increases until it reaches a maximum at an inclination of 43°. As the inclination increases above 43°, the number of daylight passes decreases.

For a satellite in a circular orbit at an altitude of 350 km and with a minimum elevation angle requirement of 10°, the effective Earth central angle is approximately 11° as depicted in the next figure. For this case the optimum inclination occurs at 10° above the target latitude.

12.9.5 Constellation Design

Figure shows the distribution of passes made over a 30 day period for a satellite in a 500 km circular orbit with a minimum elevation angle of 10° and a target at a latitude of 33°. On each day the passes are made in succession with approximately 95 minutes, the period of the orbit, between passes. The number of passes made per day ranges from a maximum of 6 to a minimum of 4.

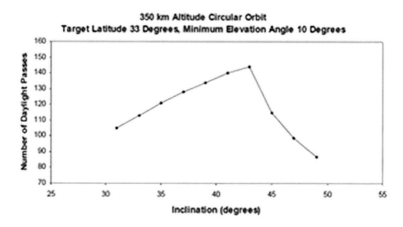

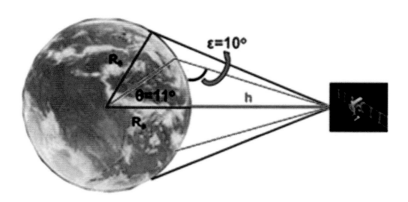

12.9.5.1   Distribution of Satellite Passes to 500 km Altitude Circular Orbit, 45 Inclination, Target Latitude 33, Minimum Elevation Angle 10

The time of day when the passes occur drifts as the orbit's node regresses. At the beginning of the time period, the passes occur later in the day but on subsequent days the passes occur at an earlier time of day until at the end of the 30 day period the passes are occurring in the early portion of the day. The impact of this trend is that there are fewer passes on days near the beginning and end of the time period.

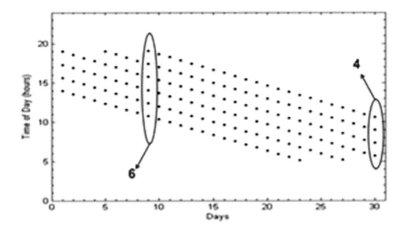

Two Satellites Separated by 180° in Mean Anomaly

One constellation design is to place two satellites in the same orbit but separated by mean anomaly. Next figure shows the distribution of satellite passes for two satellites separated by 180° of mean anomaly. Both satellites were in 500 km altitude circular orbits at an inclination of 45° and with a longitude of the ascending node of 72°. A minimum elevation angle of 10° was used and the target was located at a latitude of 33°. The number of passes per day ranges from a maximum of 12 to a minimum of 7. The passes each day occur successively with approximately 47 minutes between each pass. The number of passes has nearly doubled from 159 passes for one satellite to 316 passes with the additional satellite. Similar to the one satellite case, there are fewer passes on days at the beginning and end of the 30 day period [7].

12.9.5.2 Distribution of Passes for Two Satellites Over 30 Days Satellite 1 and 2: 500 km Altitude Circular Orbit, 45 Inclination, 72 Longitude of Ascending Node, Minimum Elevation Angle 10, Target Latitude 33

12.10   Conclusions

12.10.1   Target Location

The latitude at which a target site is located plays an important role in determining the appropriate orbit for target coverage. If the target site is on the equator, an equatorial orbit should be used. In this case the orbit's inclination will match the target site's latitude. In an equatorial orbit, the satellite's field of view will always be

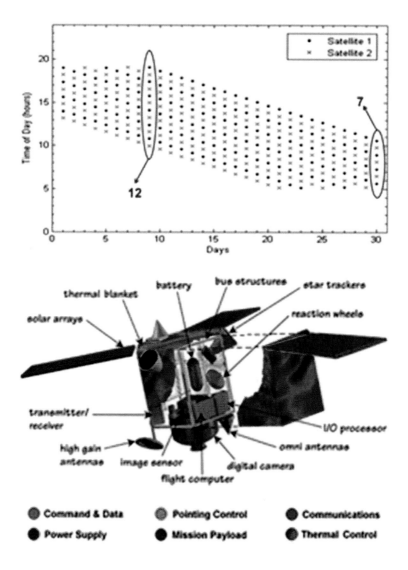

over the equator and whenever the target site's longitude comes within the satellite's field of view, the target site will be visible to the satellite. No additional gains are made by increasing the orbital inclination above the equator. If the inclination is increased a few degrees but remains below the value of the Earth-central angle for a given altitude, the equator will still always be within the field of view of the satellite. However the average slant range from the satellite to the target will be increased which is undesirable for applications such as high-resolution visible imaging. If the inclination is increased above the value of the Earth-central angle, the latitude of the target site will no longer always be within the field of view of the satellite and the number of daylight passes made will decrease. If the target site is at a latitude that is above the equator but at a smaller value than the Earth-central angle, the case is slightly different. The altitude of the orbit will determine the Earth-central angle for an altitude range of 200 to 800 km the Earth central angles range from 14 to 27°. Thus the latitude range that this case applies to will vary depending on the orbital altitude, but low latitudes such as 5 or 10° will always fall in this category. If the satellite is placed in an equatorial orbit, the latitude of the target site will always be within the satellite's field of view because the latitude value is less than the Earth-central angle. The satellite's inclination can be raised above the equator and the latitude of the target site still always be within the field of view. This will be the case if the inclination selected is less than the difference between the Earth-central angle and the target latitude. For example the 350 km altitude orbit has an Earth-central angle of about 18.6°. For a target site at a latitude of 10°, the difference between the Earth-central angle and target latitude is about 8.6°. At an inclination above 8.6°, the target latitude would no longer always be within the field of view of the satellite which would mean a decrease in the number of passes. The results for the 350 km case showed a decrease in the number of passes at inclinations above 8°. Since the number of passes does not vary significantly for the range of inclinations in which the target latitude is always within the flied of view of the satellite, a simple solution is to just choose an equatorial orbit. However the average slant range will be minimized at an inclination near the target latitude. In the inclination range where the latitude is always within the field of view, the average slant range might be higher at an equatorial orbit than an inclination closer to the target latitude. The trade-offs between slant range and number of passes will have to be considered before choosing an orbit. Another case is when the Earth-central angle is less than the value of the target latitude. Depending on the orbit altitude this range of latitude would begin around 18–27°. Since most recent theater operations have occurred at latitudes above 27°, this region is of high interest. There are two types of coverage that can be provided by low altitude orbits for targets in this latitude region. The first type of coverage is when a satellite makes one pass over the latitude of the target during each orbital period. This type of coverage will include the range of inclinations from the latitude of the target up to a value near the Earth-central angle plus the latitude of the target. The second is when the satellite makes two passes, one as it is ascending and one as it is descending, over the latitude of the target. This range will include inclinations higher than the Earth-central angle plus the latitude of the target. The highest amount of daylight passes will be made by the first type of coverage and will occur at an inclination near the value of the latitude plus the Earth-central angle. The results have shown the second case to be undesirable because the number of daylight passes decreases significantly. The average slant range also decreases but not enough to compare with the first type of coverage. For the first type of coverage, as the inclination increases above the target latitude, the number of daylight passes increases but so does the average slant range. The trade-off between slant range and passes is an important consideration for orbit selection [11].

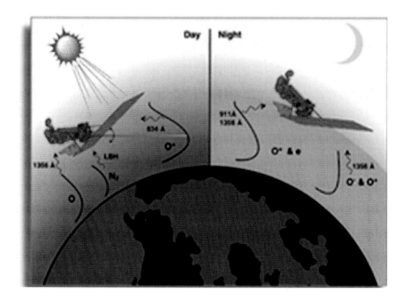

12.10.2   Orbital Altitude

The altitude of the orbit is another important parameter that affects the target coverage a satellite provides. For a target site located on the equator and a satellite in an equatorial orbit, increasing the altitude will decrease the number of daylight passes made over the target. This trend corresponds with the increased period a higher altitude orbit will have. The period of the orbit is important because during the time it takes the satellite to complete an orbit, the Earth will rotate. For a higher period, there will be fewer successive passes because the Earth will rotate more during each orbit than for a shorter period and the target will be out of view after less passes than for a shorter period. Increasing the orbit altitude also increases the field of view and hence Earth-central angle of the satellite but since the latitude of the target is always within view, a larger field of view does not add any increase in the number of passes. Increasing the altitude of the orbit also increases the average slant range to the target. Since the satellite is at a higher altitude, the distance from the satellite to the target will also be higher. If the target site is at a latitude of $10°$, the effects of increasing the orbit altitude are similar to the equatorial case. At a given altitude the number of daylight passes and average slant range vary with inclination, but there are still overall trends that are evident for varying altitudes. As the orbit altitude increases, the period of the orbit increases and the number of daylight passes decreases. The average slant range also increases as the orbit altitude increases. For a target latitude of $33°$, orbit altitude has several important effects. At a particular altitude the number of daylight passes and average slant range will depend on inclination, but there are still general trends that can be observed for varying altitudes. As the altitude of an orbit is increased, the Earth-central angle of the satellite is increased. This corresponds to an increase in the number of successive passes that are made by a satellite. If the orbit altitude is increased, the number of daylight passes made increases. The average slant range also increases with increasing orbit altitude [8,9].

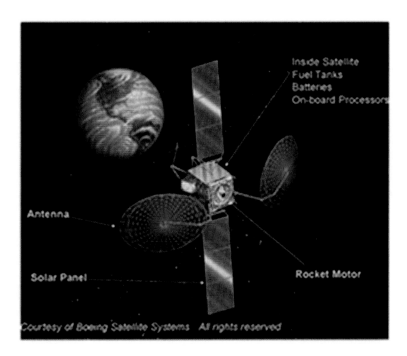

12.11   Recommendations for Future Work

The satellite propagation included the J2 perturbation which causes a regression of the node for an orbit. The perturbation was included because it has an important impact on orbit coverage. Another parameter that could be included is the drag force on a satellite. Satellites in low Earth orbit experience a significant force due to drag which could be modeled in order to see its impact on target coverage. The orbit optimization algorithm included two important coverage properties, the number of daylight passes and the average slant range. There are various other coverage properties that could be included in an optimization algorithm. The algorithm could include the objectives of maximizing the total coverage time over the target or minimizing the average or maximum time between passes. An algorithm could also be developed to optimize constellations of satellites. The focus of this research effort was on orbits which will be used for satellites collecting visible imagery.

Tactical satellites may also serve other missions such as communications or types of surveillance other than visible imaging. Since other applications may be able to operate at night, the requirement for daylight passes would not necessarily be included. Other requirements such as a minimum time between passes could be explored [13].

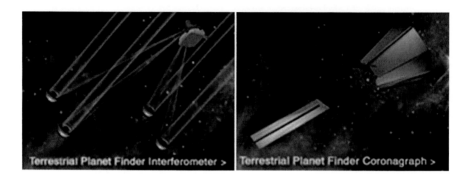

References

1. Beste, D.C.: Design of satellite constellations for optimal continuous coverage. IEEE Trans. Aero. Electron. Syst. **AES-14**(3), 466–473 (1978)
2. Cartwright, J.E.: Assured access to space. High Frontier **3**(1), 3–5 (2006)
3. Draim, J.E.: Lightsat constellation designs. AIAA Satellite Communications Conference, Washington DC, pp. 1361–1369 (1992)
4. Emery, J.D. et al.: The utility and logistics impact of small-satellite constellations in matched inclination orbits. MS thesis, AFIT/GSE/ENY/05-M01. Graduate School Of Management and Engineering, Air Force Institute of Technology (AU), Wright-Patterson AFB OH (2005)
5. Hanson, J.M., Maria, J.E., Ronald, E.T.: Designing good partial coverage satellite constellations. AIAA/AAS Astrodynamics Conference, Portland, OR, pp. 214–231, August 1990
6. Lang, T.J.: Low earth orbit satellite constellations for continuous coverage of the mid-latitudes. AIAA/AAS Astrodynamics Conference, San Diego, CA, pp. 595–607, July 1996
7. Lang, T.J.: Optimal low earth orbit constellations for continuous global coverage. Proceedings of the AAS/AIAA Conference, Victoria, Canada, pp. 1199–1216, August 1993
8. Lang, T.J.: Orbital constellations which minimize revisit time. Proceedings of the AAS/AIAA Conference, Lake Placid, NY, San Diego, pp. 1071–1086, August 1983
9. Rendon, A.: Optimal coverage of theater targets with small satellite constellations. MS thesis, AFIT/GSS/ENY/06-M12, Graduate School Of Management and Engineering, Air Force Institute of Technology (AU), Wright-Patterson AFB OH, March 2006
10. Vallado, D.A.: Fundamentals of Astrodynamics and Applications. 2nd edn. Microcosm, El Segundo (2001)
11. Walker, J.G.: Satellite constellations. JBIS **37**, 559–571 (1984)
12. Wertz, J.R.: Coverage, responsiveness, and accessibility for various responsive orbits, 3rd Responsive Space Conference, Los Angeles, CA, April 2005, Microcosm, El Segundo, 2005
13. Wertz, J.R., Wiley, J.L.: Space Mission Analysis and Design. 3rd edn. Microcosm, Torrance (1999)

Part VII
Coding, Statistical Modelling
and Applications

Chapter 13
A Bird's-Eye View of Modern Symmetric Cryptography from Combinatorial Designs

Christos Koukouvinos and Dimitris E. Simos

Abstract In the past few decades, combinatorial design theory has grown to encompass a wider variety of investigations, many of which are not apparently motivated by any practical application. Rather, they are motivated by a desire to obtain a coherent and powerful theory of existence and properties of designs. Nevertheless, it comes as no surprise that applications in coding theory and communications continue to arise, and also that designs have found applications in new areas. Cryptography in particular has provided a new source of applications of designs, and simultaneously a field of new and challenging problems in design theory.

In this paper, we present a number of applications of combinatorial designs in which the connection with modern symmetric (private-key) cryptography appears to be substantial and meaningful. We survey recent powerful private-key cryptosystems from special classes of combinatorial designs, i.e., orthogonal and Plotkin arrays, Hadamard matrices which are constructed from one and two circulant cores, which possess beautiful combinatorial properties. In addition, we present a new symmetric cryptosystem based on the famous Williamson construction for Hadamard matrices. Practical aspects of the cryptosystems, in terms of security and cryptanalysis, are analyzed and examples of real-time encryption and decryption are provided using cryptographic algorithms. We conclude by providing a state-of-the-art comparison of private-key block ciphers in the field of modern cryptography.

Keywords Encryption • block ciphers • combinatorial designs

Mathematics Subject Classification (2010): 05B20, 68P25, 94A60

C. Koukouvinos • D.E. Simos (✉)
Department of Mathematics, National Technical University of Athens,
Zografou 15773, Athens, Greece
e-mail: ckoukouv@math.ntua.gr; dsimos@math.ntua.gr

N.J. Daras (ed.), *Applications of Mathematics and Informatics in Military Science*,
Springer Optimization and Its Applications 71, DOI 10.1007/978-1-4614-4109-0_13,
© Springer Science+Business Media New York 2012

13.1 Introduction

In this paper, we survey recent private symmetric key ciphers based on several constructions that have arisen using binary arrays of combinatorial designs. In addition, we present a new symmetric cryptosystem for a specific class of combinatorial designs called Williamson matrices. By the term symmetric we mean that the same key is used both for encryption and decryption of a message. The respective cryptographic algorithms for the encryption and decryption process are called symmetric key block ciphers, and divide the original message which is going to be encrypted into blocks and encrypt each block separately. In this family of ciphers, the motivation for using Hadamard and Williamson Matrices, and orthogonal and Plotkin arrays was that these designs are often hard to find and the algorithms for encryption and decryption are of reasonable length. For encryption methods based on combinatorial designs we refer the interested reader to [37]. Applications of combinatorial designs to communications, cryptography, and networking can be found in the survey paper, [8].

13.1.1 Specifications

The cipher has similarities to the Hill cipher, i.e., using the incidence matrix of a combinatorial design for encryption and decryption and to the one time pad [31,46]. For more details regarding the Hill encryption method, see [46]. Moreover, we present a unified approach for iterated versions of these combinatorial design ciphers through the use of Kronecker product that approximates a k-round Feistel cipher or network [31]. Widely known ciphers that use the block structure of a Feistel network are data encryption standard (DES), Blowfish [38], FEAL [43], and the LOKI family of ciphers (LOKI89, LOKI91, [6]). A list of typical attacks and reference of the existing protocols can be found in ([12] and [5]), respectively. The design goals set for the combinatorial design ciphers include the following:

1. Include randomness in the encryption process
2. Require the key be shared only once
3. Use a relatively small key size
4. Computationally fast
5. The ciphers have good diffusion
6. Robust to most common cryptographic attacks

The ciphers we survey in this paper implement goals 2, 3, and 4. We shall illustrate that one variant of them (Plotkin ciphers) includes randomness in the encryption process and demonstrate that all the ciphers have good diffusion properties and provide resistance to most common cryptographic attacks.

The encryption process can be described from the following procedure: consider a communication channel, we divide the channel into two subbands, one which will carry the message and the other which will carry noise. The message, along with the noise, is transmitted over the channel. The recipient then filters out the noise, leaving only the message. This procedure is carried out using combinatorial designs.

This paper can be regarded as a unification and expansion of the proposed schemes given in [24–26], and it is organized as follows. In Sect. 13.2, we present the cryptographic algorithms used for all the encryption schemes. In Sect. 13.3 we design the encryption schemes using combinatorial designs, while in Sect. 13.4 we consider practical aspects of the proposed ciphers. Finally in Sect. 13.5 we study the security of all present private-key cryptosystems and conclude with a comparison of modern symmetric block ciphers in the field of cryptography.

13.2 Cryptographic Algorithms

We assume that the message to be transmitted is a plaintext with n letters, which is represented by a vector of length n, whereas each coordinate of the vector is a numerical value of the corresponding letter of the plaintext (i.e., ASCII code). We note that the design of cryptographic algorithms given here are a generalization of the ones given in [24], since in this paper we explore the use of orthogonal matrices generated by combinatorial designs instead of orthogonal arrays.

If the message has more than n letters then the procedure which is given below is being repeated as much times as needed. If it has less than n letters then we pad the plaintext with the letter "space" sufficient times. For the requirements of the proposed encryption method we will make use of a matrix A of order $n \times n$, of special structure, with entries $\{\pm 1\}$ where the matrix A satisfies $AA^T = kI_n$ for some constant $k \in IN$, where T stands for transposition and I_n is the identity matrix of order n. Design theory is rich of such matrices of special structure having beautiful combinatorial properties, i.e., Hadamard matrices. For more details on the application of combinatorial designs in cryptography we refer the interested reader to [8, 37].

If the message we wish to transmit has been converted to a numerical vector $\bar{m}$, then the encrypted message which is going to be transmitted over a communication channel is

$$\bar{c} = \bar{m}A + d\bar{e}_n$$

where d is a suitable constant and $\bar{e}_n = (1, \ldots, 1)$ is a $1 \times n$ vector of ones. The receiver in order to decrypt the encrypt message has to make use of the transformation $\bar{m} = 1/k(\bar{c} - d\bar{e}_n)A^T$, where A^T is the transpose of the matrix A which has been used during the encryption. The encryption method described previously can be implemented with the following cryptographic algorithm given in [26].

In order for the encryption method to be persistent with respect to the basic cryptographic principles, the encrypted message $\bar{c}$ has to be decrypted uniquely. This requirement is satisfied from the following theorem.

Algorithm 1 Encryption Algorithm

---

function ENCRALG(msg)
Require: msg in ASCII code   ▷ Encode a sample plaintext, msg
   SELECT(A,d)   ▷ Choose appropriate A and d
   $k \leftarrow (A,d)$ ▷ Form private key k
   TRANSMIT(k)   ▷ Transmit securely the private key
   $\bar{m} \leftarrow$ CONVERT(msg)   ▷ Convert original msg
   $\bar{c} \leftarrow \bar{m}A + d\bar{e}_n$ ▷ Encrypted msg is $\bar{c}$
   return (TRANSMIT($\bar{c}$))
end function

Theorem 13.1 (Koukouvinos and Simos [26]). *The encrypted message $\bar{c}$ which is transmitted with respect to the encryption algorithm is decrypted uniquely as $\bar{w} = 1/k(\bar{c} - d\bar{e}_n)A^T$ and $\bar{w} \equiv \bar{m}$.*

The decryption process uses the previous theorem as its cornerstone and is implemented with the following cryptographic algorithm, again given in [26].

Algorithm 2 Decryption Algorithm

---

function DECRALG($\bar{c}$)
Require: given ciphertext $\bar{c}$ ▷ Decode a given ciphertext
   RECEIVE(A,d)   ▷ Receive the securely transmitted private key
   $k \leftarrow (A,d)$ ▷ Set private key k
   $\bar{m} \leftarrow 1/k(\bar{c} - d\bar{e}_n)A^T$ ▷ Decrypt ciphertext $\bar{c}$
   $msg \leftarrow$ CONVERT($\bar{m}$)   ▷ Original plaintext is msg
   return (msg)
end function

13.3   Private-key Ciphers

In this section, we provide several constructions for encryption schemes using one array of special structure. We give some necessary notations and definitions that we shall use throughout this paper. We note that all arrays that are used below can be considered as binary array bits with the aid of the following $\{1,-1\}$-bit notation taken from [28].

Definition 13.2 ($\{1,-1\}$-bit Notation). Sometimes, we find it convenient to view bits as being $\{1,-1\}$-valued instead of $\{0,1\}$-valued. If $b \in \{0,1\}$ then $\bar{b} \in \{1,-1\}$ is defined to be $\bar{b} = (-1)^b$. If $x \in \{0,1\}^n$ then $\bar{x} \in \{1,-1\}^n$ is defined as the string where the ith bit is $\bar{x}_i$.

A cipher's strength is determined by the computational power needed to break it. The computational complexity of an algorithm is measured by two variables: T for time complexity which specifies how the running time depends on the size of the input and S for space complexity or memory requirement. Both T and S are commonly expressed as functions of n, when n is the size of the input.

Generally, the computational complexity of an algorithm is expressed in what is called "big $\mathcal{O}$" notation; the order of magnitude of the computational complexity. We use $\mathcal{O}$-notation to give an upper bound on a function, to within a constant factor [9].

Definition 13.3 ($\mathcal{O}$-Notation). For a given function $g(n)$ we denote by $O(g(n))$ the set of functions $O(g(n)) = \{f(n): \text{ there exist positive constants } c \text{ and } n_0 \text{ such that } 0 \le f(n) \le cg(n) \text{ for all } n \ge n_0\}$.

We give a necessary brief definition for an encryption scheme.

Definition 13.4 (Boyd and Mathuria [5]). An encryption scheme consists of three sets: a key set K, a message set M, and a ciphertext set C together with the following three algorithms.

1. A key generation algorithm, which outputs a valid encryption key $k \in K$ and a valid decryption key $k^{-1} \in K$.
2. An encryption algorithm, which takes an element $m \in M$ and an encryption key $k \in K$ and outputs an element $c \in C$ defined as $c = E_k(m)$.
3. A decryption function, which takes an element $c \in C$ and a decryption key $k^{-1} \in K$ and outputs an element $m \in M$ defined as $m = D_k^{-1}(c)$. We require that $D_k^{-1}(E_k(m)) = m$.

Remark 13.5. We note that although we have used as a private key the pair (A, d), in terms of computational complexity, henceforth we can refer to the private key using only the encryption matrix A since d is of size $\mathcal{O}(1)$.

It is clear that since we have an encryption algorithm and a decryption function we need a key generation algorithm in order to construct an encryption scheme. This key generation algorithm will be derived each time from a class of combinatorial designs; thus in the following sections, we name the ciphers after the respective combinatorial structure used.

13.3.1 OA Ciphers

Orthogonal arrays were introduced by Rao [35, 36] over half a century ago. Applications of them have arisen in many areas of discrete mathematics and statistics. For further details on orthogonal arrays we refer the interest reader to [20].

Definition 13.6. An orthogonal array $OA(n,q,s,t)$ is an $n \times q$ array with entries from a set of s distinct symbols arranged so that, for any collection of t columns of the array, each of the s^t row vectors appears equally often.

In the case of $s = 2$ we can use as the set of the symbols the set $S = \{-1,1\}$. Then for any $t \geq 2$ (we will always think and use t as 2), by definition, it is easy to verify that for any selection of two distinct columns of an orthogonal array the usual inner product of the columns is zero. As an encryption matrix we will use the transpose of an orthogonal array with parameters $(n,q,2,t)$.

Definition 13.7. Two orthogonal arrays based on s symbols are said to be *isomorphic* if one can be obtained from the other by a sequence of row permutations, column permutations, and permutations of symbols in each column.

It is a hard combinatorial problem to find a complete set of non-isomorphic orthogonal arrays with certain parameters, [1,7,11]. Also much work has been done towards the enumeration of non-isomorphic orthogonal arrays [48].

Example 13.8. Below we present an orthogonal array: $OA(12,4,2,2)$

$$
\begin{bmatrix}
1 & 1 & 1 & 1 \\
1 & 1 & 1 & 1 \\
1 & 1 & -1 & -1 \\
1 & -1 & 1 & -1 \\
1 & -1 & -1 & 1 \\
1 & -1 & -1 & -1 \\
-1 & 1 & 1 & -1 \\
-1 & 1 & -1 & 1 \\
-1 & 1 & -1 & -1 \\
-1 & -1 & 1 & 1 \\
-1 & -1 & 1 & -1 \\
-1 & -1 & -1 & 1
\end{bmatrix}
$$

In this case the key k will be the orthogonal array, A, $OA(n,q,2,t)$, which consists by $n \times q$ bits. In terms of computational complexity, since $q < n$, the size of the key is of order $\mathcal{O}(n^2)$.

Proposition 13.9 (Koukouvinos et al. [24]). *There exist a family of private-key ciphers using orthogonal arrays with parameters $OA(n,q,2,t)$, $t \geq 2$, which will be called OA ciphers.*

Remark 13.10. It is obvious that the use of two non-isomorphic or different (maybe isomorphic) orthogonal arrays with the same parameters will result in two different ciphertexts.

Theorem 13.11 (Hedayat et al. [20]). *For every n multiple of 4 there exists a two-level orthogonal array with parameters $OA(n,q,2,t)$ for some q and $t \geq 2$.*

13.3.2   Hadamard Ciphers

Hadamard matrices are named after Jacques Hadamard, who found square matrices of orders 12 and 20, with entries ± 1, which had all their rows (and columns) orthogonal [18].

Definition 13.12. A *Hadamard matrix* of order n is a square $n \times n$ matrix H whose elements are $+1$'s and -1's, with the property

$$HH^T = nI_n$$

where T stands for transposition and I_n is the identity matrix of order n.

The Hadamard property entails that the rows (and columns) of a Hadamard matrix are pairwise orthogonal. It is well known that if n is the order of a Hadamard matrix then n is necessarily $1, 2$ or a multiple of 4. Hadamard matrices are used in Combinatorics, Statistics, Coding Theory, Telecommunications, and other areas. More details on Hadamard matrices can be found in [10, 42].

As an encryption matrix for this scheme we will use a Hadamard matrix of order n. In the case of Hadamard matrices it is obvious that the use of two different Hadamard matrices of the same order will result in two different ciphertexts, due to the presence of the H-equivalence property described below.

Two Hadamard matrices are called equivalent (or Hadamard equivalent or H-equivalent) if one can be obtained from the other by a sequence of row negations, row permutations, column negations, and column permutations. More specifically, two Hadamard matrices are equivalent if one can be obtained by the other by a sequence of the following transformations:

- Multiply rows and/or columns by -1.
- Interchange rows and/or columns.

Two Hadamard matrices are called inequivalent, if they are not equivalent. Therefore, the choice of inequivalent Hadamard matrices as encryption matrices ensures that two inequivalent Hadamard matrices will result in two different ciphertexts. Otherwise one could transform the one encryption matrix to another, following the transformations mentioned above.

It is vital for our application to have large databases of inequivalent matrices to our disposal. As of release 2.13, Magma contains a database of inequivalent Hadamard matrices. There exist several thousands (even millions) of inequivalent Hadamard matrices for some orders. As an example for order 32 which is a reasonable length for the encryption process there are more than $3,578,006$ inequivalent Hadamard matrices [32].

The private key k used in the encryption process will be the Hadamard matrix of order n, $A = H_n$, which consists of $n \times n$ bits. In terms of computational complexity, the size of the key is $\mathcal{O}(n^2)$.

Proposition 13.13 (Koukouvinos and Simos [26]). *There exist a family of private-key ciphers using Hadamard matrices of order n, which will be called Hadamard ciphers.*

There are some special constructions of Hadamard matrices which enable us to reduce the size complexity of the private key.

13.3.2.1 Hadamard Core Ciphers

A Hadamard matrix of order $p+1$ which can be written in one of the two equivalent forms

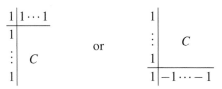

or

where $C = (c_{ij})$ is a circulant matrix of order p, i.e., $c_{ij} = c_{1,j-i+1 (\mathrm{mod}\ p)}$ is said to have a circulant core. The following matrices are examples for order 12.

1	1	1	1	1	1	1	1	1	1	1	1
1	-	1	-	1	1	1	-	-	-	1	-
1	-	-	1	-	1	1	1	-	-	-	1
1	1	-	-	1	-	1	1	1	-	-	-
1	-	1	-	-	1	-	1	1	1	-	-
1	-	-	1	-	-	1	-	1	1	1	-
1	-	-	-	1	-	-	1	-	1	1	1
1	1	-	-	-	1	-	-	1	-	1	1
1	1	1	-	-	-	1	-	-	1	-	1
1	1	1	1	-	-	-	1	-	-	1	-
1	-	1	1	1	-	-	-	1	-	-	1
1	1	-	1	1	1	-	-	-	1	-	-

1	1	-	1	-	-	-	1	1	1	-	1
1	1	1	-	1	-	-	-	1	1	1	-
1	-	1	1	-	1	-	-	-	1	1	1
1	1	-	1	1	-	1	-	-	-	1	1
1	1	1	-	1	1	-	1	-	-	-	1
1	1	1	1	-	1	1	-	1	-	-	-
1	-	1	1	1	-	1	1	-	1	-	-
1	-	-	1	1	1	-	1	1	-	1	-
1	-	-	-	1	1	1	-	1	1	-	1
1	1	-	-	-	1	1	1	-	1	1	-
1	-	1	-	-	-	1	1	1	-	1	1

1	-	-	-	-	-	-	-	-	-	-	-

Where $-$ stands for -1 to conform with the customary notation for Hadamard matrices. The two forms are equivalent as described earlier.

The scheme is constructed by using the previous Hadamard matrix $A = H_n$ of order $n = 4m = p + 1$ as an encryption matrix. However, in this case the circulant structure of the Hadamard matrix gives us the opportunity to use a key of a significant less size than previously as follows.

Let $A_c = [a_1, a_2, \ldots, a_p]$ denote the first row of the circulant matrix, C, used in the one circulant core construction previously. The private key k for this scheme is the binary vector, A_c, which consists of p bits. Therefore, when a Hadamard matrix of order $n = p + 1$ is used as an encryption matrix the key is of size $\mathcal{O}(n)$, since it consists of $p = n - 1$ bits.

Proposition 13.14 (Koukouvinos and Simos [26]). *There exist a family of private-key ciphers using Hadamard matrices with one circulant core of order $n = p + 1$, which will be called Hadamard core ciphers.*

Four families of these kinds of Hadamard matrices have been found by Paley [33], Stanton, Sprott and Whiteman [45, 54], Singer [44] and Marshall Hall [19], which can be used in the previous proposition and give rise to infinite families of encryption schemes based on Hadamard matrices with one circulant core. The following theorem was given in [21].

Theorem 13.15 (Circulant Core Hadamard Construction Theorem). *A Hadamard matrix of order $p + 1$ with circulant core can be constructed if*

1. $p \equiv 3 \pmod 4$ *is a prime [33]*
2. $p = q(q + 2)$ *where q and $q + 2$ are both primes [45, 54]*
3. $p = 2^t - 1$ *where t is a positive integer [44]*
4. $p = 4x^2 + 27$ *where p is a prime and x a positive integer [19]*

13.3.2.2 Hadamard Cores Ciphers

A Hadamard matrix of order $2\ell + 2$ (for ℓ odd) which can be written in one of the two equivalent forms ($-$ stands for -1 and $+$ stands for $+1$)

$$
\begin{bmatrix}
- & - & + & \cdots & + & + & \cdots & + \\
- & + & + & \cdots & + & - & \cdots & - \\
+ & + \\
\vdots & \vdots & & A & & & B \\
+ & + \\
\hline
+ & - \\
\vdots & \vdots & & B^T & & & -A^T \\
+ & -
\end{bmatrix}
\quad \text{or} \quad
\begin{bmatrix}
+ & + \\
\vdots & & A & & & B \\
+ & + \\
+ & - \\
\vdots & & B^T & & & -A^T \\
+ & - \\
\hline
- & - & + & \cdots & + & + & \cdots & + \\
- & + & + & \cdots & + & - & \cdots & -
\end{bmatrix}
$$

where $A = (a_{ij})$, $B = (b_{ij})$ are two circulant matrices (with ± 1 elements) of order ℓ, i.e., $a_{ij} = a_{1,j-i+1 (\mathrm{mod}\ \ell)}$, $b_{ij} = b_{1,j-i+1 (\mathrm{mod}\ \ell)}$, is said to have two circulant cores.

As before the scheme is constructed by using the previous Hadamard matrix $A = H_n$ of order $n = 2\ell + 2$ as an encryption matrix. However, in this case the circulant structure of the Hadamard matrix gives us the opportunity to use a key of a significant less size than previously as follows.

Let $A_c = [a_1, a_2, \ldots, a_\ell]$ and $B_c = [b_1, b_2, \ldots, b_\ell]$ denote the first row of the circulant matrices, A and B, used in the two circulant core construction, respectively. The private key k for this scheme is the concatenation of the two vectors, A_c and B_c, denoted by $A_c \oplus B_c$ which consists of $\ell + \ell$ bits. Therefore, when a Hadamard matrix of order $n = 2\ell + 2$ is used as an encryption matrix the key is of size $\mathcal{O}(n)$, since it consists of $2\ell = n - 2$ bits.

Proposition 13.16 (Koukouvinos and Simos [26]). *There exist a family of private-key ciphers using Hadamard matrices with two circulant cores of order $n = 2\ell + 2$, which will be called Hadamard cores ciphers.*

Since $2\ell + 2$ must be equal to a multiple of 4 we have that ℓ must be an odd integer for this construction to yield a Hadamard matrix.

Georgiou, Koukouvinos and Seberry [15] point out that *GL*-pairs, which can be used to construct Hadamard matrices of order $2\ell + 2$ with two circulant cores, exist for many cases. These matrices can be used in the previous proposition and give rise to infinite families of encryption schemes based on Hadamard matrices with two circulant cores. The following theorem was given in [22].

Theorem 13.17 (Two Circulant Cores Hadamard Construction Theorem). *A Hadamard matrix of order $2\ell + 2$ with with two circulant cores can be constructed if*

1. *ℓ is a prime (see, for example, [13])*
2. *$2\ell + 1$ is a prime power (these arise from Szekeres difference sets, see, for example, [13] or [16])*
3. *$\ell = 2^k - 1$, $k \geq 2$ (two Galois sequences are a GL-pair, see, for example, [40])*
4. *$\ell = p(p+2)$ where p and $p+2$ are both primes (two such sequences are a GL-pair, see, for example, [45, 54])*
5. *$\ell = 49, 57$ (these have been found by a non-exhaustive computer search that uses generalized cyclotomy and master-switch techniques, see [16, 17])*
6. *$\ell = 3, 5, \ldots, 45$ (these have been found and classified by exhaustive computer searches, see [13])*
7. *$\ell = 47, 49, 51, 53$, and 55 (these have been found and classified by partial computer searches, see [13])*
8. *$\ell = 143$ (also verified the results for $\ell = 3, 5, 7, 11, 13, 15, 17, 19, 23, 25, 31, 35, 37, 41, 43, 53, 59, 61, 63$ see [14])*

13.3.3 Plotkin Ciphers

Definition 13.18. An *orthogonal design* of order n and type $(s_1, s_2, \ldots, s_k)$ denoted $OD(n; s_1, s_2, \ldots, s_k)$ in the commuting variables $x_1, x_2, \ldots, x_k$ is a square matrix D of order n with entries from the set $\{0, \pm x_1, \pm x_2, \ldots, \pm x_k\}$ satisfying

$$DD^T = \sum_{i=1}^{k} (s_i x_i^2) I_n,$$

where I_n is the identity matrix of order n.

Orthogonal designs are used in Combinatorics, Statistics, Coding Theory, Telecommunications and other areas. More details on orthogonal designs and Hadamard matrices can be found in [41, 42]. The last definitions give us the following insights:

1. In any row there are s_1 entries $\pm x_1$, s_2 entries $\pm x_2, \ldots, s_k$ entries $\pm x_k$, and similarly for the columns.
2. The rows and columns are pairwise orthogonal, respectively.

The choice of orthogonal designs for constructing orthogonal matrices and afterwards encryption schemes enable us to choose between a large variety of classes of orthogonal designs with different structure. Plotkin [34] showed that, if there is a Hadamard matrix of order $2t$, then there is an $OD(8t; t, t, t, t, t, t, t, t)$. It is conjectured that there is an $OD(8n; n, n, n, n, n, n, n, n)$ for each odd integer n. These orthogonal designs are called *Plotkin arrays*.

We initiate the construction of the encryption scheme based on Plotkin arrays. Note that this cipher implements in addition the design goal of adding randomness during the encryption process. As an example, we illustrate the construction based on the Plotkin array of order 8 and type $(1, 1, 1, 1, 1, 1, 1, 1)$. The corresponding orthogonal design is the following:

$$OD(8; 1, 1, 1, 1, 1, 1, 1, 1) = \begin{pmatrix} A & B & C & D & E & F & G & H \\ -B & A & D & -C & F & -E & -H & G \\ -C & -D & A & B & G & H & -E & -F \\ -D & C & -B & A & H & -G & F & -E \\ -E & -F & -G & -H & A & B & C & D \\ -F & E & -H & G & -B & A & -D & C \\ -G & H & E & -F & -C & D & A & -B \\ -H & -G & F & E & -D & -C & B & A \end{pmatrix}, \quad (13.1)$$

If we call the above matrix P, we have that $PP^T = f I_8$ whereas $f = A^2 + B^2 + \cdots + H^2$. The Plotkin arrays allow easy construction of matrices needed in our encryption schemes. For the encryption process we have only to compute the matrix P. The encryption process starts with a message m of arbitrary length, and dividing m into blocks $m_1, \ldots, m_q$ of length 4 (padding the last block with zeros if necessary). Then random vectors $g_1, \ldots, g_q$ of length 4 are chosen. For the construction of noise vectors $g_1, \ldots, g_q$ pseudorandom generators were constructed using techniques from [28]. Finally, the matrix P is applied successively to $m_i \oplus g_i$. The ciphertext is then $c = P(m_1 \oplus g_1) \oplus \cdots \oplus P(m_q \oplus g_q)$. The notation $m \oplus g$ means that m is concatenated with g.

The message is then decrypted by dividing c into blocks $c_1, \ldots, c_q$ of size 8, computing $P^T c_i / f$ for $i = 1, \ldots, q$ and reconstructing the message using the first four entries of these blocks.

Remark 13.19. The key for the recipient is the chosen entries for P; hence in this case is the entries $A, B, \ldots, H$ of the matrix P. Therefore, the matrix P is used as an encryption matrix and the size of key is of $\mathcal{O}(n^2)$ since the matrix P consists of $n \times n$ bits.

Since the Plotkin array we used so far is relatively small, we continued by modifying appropriate the encryption process using the Plotkin array of orders 16 and 24. We note that the use of Plotkin arrays of different orders does not result in an increase to the key search space since the number of variables that appear in the aforementioned orthogonal designs remains the same. The aforementioned orthogonal designs can be found in the book [16].

Proposition 13.20 (Koukouvinos and Simos [25]). *There exist a family of private-key ciphers using Plotkin arrays of order n, which will be called Plotkin ciphers.*

13.3.4   Williamson Ciphers

In this section, we use Williamson's construction for Hadamard matrices as the basis of our construction for a new private-key symmetric cryptosystem. We briefly describe the theory of Williamson's construction below.

Theorem 13.21 (Williamson [52]). *Suppose there exist four* $(1, -1)$ *matrices A, B, C, D of order n which satisfy*

$$XY^T = YX^T, X, Y \in \{A, B, C, D\}$$

Further, suppose

$$AA^T + BB^T + CC^T + DD^T = 4nI_n \tag{13.2}$$

Then

$$H = \begin{bmatrix} A & B & C & D \\ -B & A & -D & C \\ -C & D & A & -B \\ -D & -C & B & A \end{bmatrix} \tag{13.3}$$

is a Hadamard matrix of order 4n constructed from a Williamson array.

Let the matrix T given below be called the shift matrix:

$$T = \begin{bmatrix} 0 & 1 & 0 & \cdots & 0 \\ 0 & 0 & 1 & \cdots & 0 \\ & \cdots & & \cdots & \\ 0 & 0 & 0 & \cdots & 1 \\ 1 & 0 & 0 & \cdots & 0 \end{bmatrix} \tag{13.4}$$

and note

$$T^n = I, \ (T^i)^{\mathrm{T}} = T^{n-i} \tag{13.5}$$

If n is odd, T is the matrix representation of the nth root of unity ω, $\omega^n = 1$.
Let

$$\begin{cases} A = \displaystyle\sum_{i=0}^{n-1} a_i T^i, \ a_i = \pm 1, a_{n-i} = a_i \\ B = \displaystyle\sum_{i=0}^{n-1} b_i T^i, \ b_i = \pm 1, b_{n-i} = b_i \\ C = \displaystyle\sum_{i=0}^{n-1} c_i T^i, \ c_i = \pm 1, c_{n-i} = c_i \\ D = \displaystyle\sum_{i=0}^{n-1} d_i T^i, \ d_i = \pm 1, d_{n-i} = d_i \end{cases} \tag{13.6}$$

Then matrices A, B, C, D may be represented as polynomials. The requirement that $x_{n-i} = x_i, x \in \{a, b, c, d\}$ forces the matrices A, B, C, D to be symmetric.

Since A, B, C, D are symmetric, (13.2) becomes

$$A^2 + B^2 + C^2 + D^2 = 4nI_n$$

and the relation $XY^{\mathrm{T}} = YX^{\mathrm{T}}$ becomes $XY = YX$ which is true for polynomials.

Definition 13.22. Williamson matrices are $(1, -1)$ symmetric circulant matrices. As a consequence of being symmetric and circulant they commute in pairs.

The scheme is constructed by using the Williamson Hadamard matrix $A = H_{4m}$ of order $n = 4m$ as an encryption matrix. However, in this case the circulant structure of symmetric matrices involved in the Williamson's construction gives us the opportunity to use a key of a significant less size than previously as follows.

In detail, for the encryption process is needed to construct the $(1, -1)$ circulant matrices:

$$A = [a_0, a_1, \ldots, a_{m-1}], \ B = [b_0, b_1, \ldots, b_{m-1}],$$
$$C = [c_0, c_1, \ldots, c_{m-1}], \ D = [d_0, d_1, \ldots, d_{m-1}],$$

such that

$$A^2 + B^2 + C^2 + D^2 = 4mI_m. \tag{13.7}$$

The symmetry requirement gives $v_i = v_{m-i}$, $i = 1, 2, \ldots, \frac{1}{2}(m-1)$, $v_i \in \{a_i, b_i, c_i, d_i\}$.

The private key k for this scheme is the concatenation of the four vectors, $A, B, C,$ and D, denoted by $A \oplus B \oplus C \oplus D$ which consists of $m + m + m + m$ bits. Therefore, when a Williamson Hadamard matrix of order $n = 4m$ is used as an encryption matrix the key is of size $\mathcal{O}(n)$, since it consists of $n = 4m$ bits.

Proposition 13.23. *There exist a family of private-key ciphers using Williamson Hadamard matrices of order $n = 4m$, which will be called Williamson ciphers.*

Proof. The encryption scheme using a Williamson Hadamard matrix A of order $n = 4m$, will use a key $A \oplus B \oplus C \oplus D$ of size $\mathcal{O}(n)$, as described previously, and can be encrypted–decrypted using the algorithms of Sect. 13.2 since $AA^T = nI_n$. □

An infinite family of Hadamard matrices of Williamson type has been proved to exist under certain conditions [50, 53]:

Theorem 13.24. *If q is a prime power, $q \equiv 1 \pmod 4$, $q + 1 = 2t$, then there exists a Williamson matrix of order $4t$; we have $C = D$, and A and B differ only on the main diagonal.*

This theorem gives examples of Hadamard matrices of Williamson type for orders $4t, t = 31, 37, 41, 45, 49, 51, 55, \ldots$, for example.

Results for Hadamard matrices of Williamson type can be found on the web site of C. Koukouvinos [23] and in [15]. For example, using the $\{1, -1\}$-bit notation and the four vectors $A = [1, -1, -1, -1, -1]$, $B = [1, -1, -1, -1, -1]$, $C = [1, 1, -1, -1, 1]$, and $D = [1, -1, 1, 1, -1]$ of length 5 from [23] we can construct a Williamson Hadamard matrix of order 20; which in the continuum will be used as an encryption matrix in Proposition 13.23 with a key $k = A \oplus B \oplus C \oplus D = 01111011110011001001$ of length equal to 20 bits to generate the corresponding Williamson cipher.

13.3.5 Iterated Combinatorial Design Block Ciphers

Most block ciphers are constructed by repeatedly applying a simpler function. This approach is known as iterated block cipher (or product cipher). Each iteration is termed a round, and the repeated function is termed the round function; anywhere between 4 and 32 rounds are typical. We present here a unified approach for all the combinatorial design block ciphers using Kronecker product. The product cipher will consist of a series of Kronecker products applied between the encryption matrices of the same type of the combinatorial design ciphers we have presented so far. Our goal is to achieve that the resulting cipher will be more secure than the individual components, thus making it resistant to cryptanalysis. We note that this approach shares many similarities with the design of a k-round Feistel network of ciphers.

In particular, we apply the "blow-up" construction of encryption schemes first given in [24], which relies on the previous encryption schemes and the Kronecker product as its main characteristics. We first define the Kronecker product $A \otimes B$ between two matrices A and B, a crucial definition for the construction of this family of product ciphers.

Definition 13.25 ([27]).

$$\text{Let } A = \begin{pmatrix} a_{11} & a_{12} & \dots & a_{1n} \\ \vdots & & \ddots & \\ a_{m1} & a_{m2} & \dots & a_{mn} \end{pmatrix}$$

$$\text{Then } A \otimes B := \begin{pmatrix} a_{11}B & a_{12}B & \dots & a_{1n}B \\ \vdots & & \ddots & \\ a_{m1}B & a_{m2}B & \dots & a_{mn}B \end{pmatrix}$$

If A is an $m \times n$ and B is an $p \times q$ matrix, then $A \otimes B$ is an $mp \times nq$ matrix. We note that if A and B are orthogonal matrices, then $A \otimes B$ is also an orthogonal matrix. We specialize in the case of combinatorial designs, where the round function is one use of the Kronecker product.

13.3.5.1 Kronecker OA Ciphers

Proposition 13.26 (Koukouvinos et al. [24]). *Let A, B be two $OA(n_1, q_1, 2, 2)$, $OA(n_2, q_2, 2, 2)$. Then the Kronecker product $A \otimes B$ is an orthogonal array $OA(n_1 n_2, q_1 q_2, 2, t)$, $t \geq 2$.*

Remark 13.27. We can repeat the previous construction using p orthogonal arrays $A_1, A_2, \dots, A_p$, where each A_i is an $OA(n_i, q_i, 2, 2)$ for $i = 1, \dots, p$. Thus the Kronecker product $\bigotimes_{i=1}^{p} A_i := A_1 \otimes A_2 \otimes \cdots \otimes A_p$ is an orthogonal array $OA(\prod_{i=1}^{p} n_i, \prod_{i=1}^{p} q_i, 2, 2)$.

We illustrate the construction of a Kronecker OA cipher with the following example.

Example 13.28. Let A_i, $i = 1, \dots, k$ be orthogonal arrays with parameters $(n_i, q_i, 2, 2)$. Each A_i is the private key in the case of encrypting only with A_i. If we consider the Kronecker product $\bigotimes_{i=1}^{k} A_i$ of these matrices, the generated orthogonal arrays has parameters $(\prod_{i=1}^{k} n_i, \prod_{i=1}^{k} q_i, 2, 2)$. The matrix generated by the kronecker product can be used as an encryption matrix where its private key $\bigoplus_{i=1}^{k} A_i$ is the concatenation of the private keys A_i, which consists of $\sum_{i=1}^{k} n_i q_i$ bits. We denote with n the maximum value n_i, q_i, i.e., $n = \max_i \{n_i, q_i\}$. In terms of computational complexity $\prod_{i=1}^{k} n_i q_i \leq \prod_{i=1}^{k} n^2 = n^{2k}$, the size of the encryption matrix is of exponential growth $\mathcal{O}(n^{2k})$. However, the size of the private key grows slower since $\sum_{i=1}^{k} n_i q_i < \sum_{i=1}^{k} n^2 = kn^2$; therefore its growth is of size $\mathcal{O}(n^2)$.

13.3.5.2 Kronecker Hadamard Core Ciphers

Proposition 13.29 (Sylvester [49]). *Let H_1 and H_2 be Hadamard matrices of orders m and n, respectively. Then the Kronecker product $H_1 \otimes H_2$ is a Hadamard matrix of order mn.*

Remark 13.30. We can repeat the previous construction using p Hadamard matrices $H_1, H_2, \ldots, H_p$ of orders $n_1, n_2, \ldots, n_p$. Thus the Kronecker product $\otimes_{i=1}^{p} H_i := H_1 \otimes H_2 \otimes \cdots \otimes H_p$ is a Hadamard matrix of order $\prod_{i=1}^{p} n_i$.

We illustrate the construction of a Kronecker Hadamard core cipher with the following example.

Example 13.31. Let H_i, for $i = 1, \ldots, k$ be Hadamard matrices with one circulant core of orders $n_i = p_i + 1$, for $i = 1, \ldots, k$, respectively. These matrices associated with their corresponding encryption keys $A_{c_i} = [a_{1_i}, a_{2_i}, \ldots, a_{p_i}]$ for $i = 1, \ldots, k$, where each private key A_{c_i} consists of p_i bits, form a k-family of encryption schemes or a k-round product cipher. If we consider the Kronecker product $\otimes_{i=1}^{k} H_i$ of these matrices, the generated matrix is a Hadamard matrix of order $\prod_{i=1}^{k} n_i$. Since a recipient can construct each individual Hadamard matrix H_i by assuming knowledge of the corresponding private key A_{c_i}, the matrix generated by the Kronecker product can be used as an encryption matrix where its private key $\oplus_{i=1}^{k} A_{c_i}$ is the concatenation of the private keys A_{c_i}, which consists of $\sum_{i=1}^{k} p_i$ bits. Let n denote the largest order of the Hadamard matrices we have used, i.e., $n = \max_i \{n_i\}$. In terms of computational complexity, since $\prod_{i=1}^{k} n_i \leq \prod_{i=1}^{k} n = n^k$, the size of the encryption matrix is of exponential growth $\mathcal{O}(n^k)$. However, the size of the private key grows linearly since $\sum_{i=1}^{k} p_i = \sum_{i=1}^{k} (n_i - 1) = \sum_{i=1}^{k} (n_i) - k \leq \sum_{i=1}^{k} (n) - k = kn - k = k(n-1)$; therefore its growth is of size $\mathcal{O}(n)$.

13.3.5.3 Kronecker Hadamard Cores Ciphers

Similar, we illustrate the construction of a Kronecker Hadamard cores cipher with the following example.

Example 13.32. Let H_i, for $i = 1, \ldots, k$ be Hadamard matrices with two circulant cores of orders $n_i = 2\ell_i + 2$, for $i = 1, \ldots, k$, respectively. These matrices associated with their corresponding encryption keys $A_{c_i} \oplus B_{c_i} = [a_{1_i}, a_{2_i}, \ldots, a_{\ell_i}] \oplus [b_{1_i}, b_{2_i}, \ldots, b_{\ell_i}] = [a_{1_i}, a_{2_i}, \ldots, a_{\ell_i}, b_{1_i}, b_{2_i}, \ldots, b_{\ell_i}]$ for $i = 1, \ldots, k$, where each private key $A_{c_i} \oplus B_{\ell_i}$ consists of $2\ell_i$ bits, form a k-family of encryption schemes or a k-round product cipher. If we consider the Kronecker product $\otimes_{i=1}^{k} H_i$ of these matrices, the generated matrix is a Hadamard matrix of order $\prod_{i=1}^{k} n_i$. Since a recipient can construct each individual Hadamard matrix H_i by assuming knowledge of the corresponding private key $A_{c_i} \oplus B_{c_i}$, the matrix generated by the Kronecker product can be used as an encryption matrix where its private key $\oplus_{i=1}^{k} (A_{c_i} \oplus B_{c_i})$

is the concatenation of the private keys $A_{c_i} \oplus B_{c_i}$, which consists of $\sum_{i=1}^{k} 2\ell_i = 2k \sum_{i=1}^{k} \ell_i$ bits. Let n denote the largest order of the Hadamard matrices we have used, i.e., $n = \max_i \{n_i\}$. In terms of computational complexity, since $\prod_{i=1}^{k} n_i \leq \prod_{i=1}^{k} n = n^k$, the size of the encryption matrix is of exponential growth $\mathcal{O}(n^k)$. However, the size of the private key grows linearly since $\sum_{i=1}^{k} 2\ell_i = \sum_{i=1}^{k} (n_i - 2) = \sum_{i=1}^{k} (n_i) - 2k \leq \sum_{i=1}^{k} (n) - 2k = nk - 2k = k(n - 2)$; therefore its growth is of size $\mathcal{O}(n)$.

13.3.5.4 Kronecker Plotkin Ciphers

The generation of a Kronecker Plotkin cipher can be implemented with the following algorithm.

Algorithm 3 EncoderScheme Function

---

function ENCODERSCHEME(Encodes a sample plaintext)

Step 1. Compute the encryption matrix M

Step 1a. Convert the corresponding characters of the plaintext to ASCII values.

Step 1b. Input the possible range of entries for the matrices P_i.

Step 1c. Choose the corresponding Plotkin arrays that will form the matrices P_i.

Step 1d. Compute the tensor product $M := P_1 \otimes P_2 \otimes \cdots \otimes P_p$.

Step 2. Encode the input message

Step 2a. Compute $m \oplus g$ by converting the message to ASCII values and filling the noise vector g with random numbers.

Step 2b. Compute $M(m \oplus g)$

end function

For the encryption process we choose p Plotkin arrays $P_1, P_2, \ldots, P_p$. Each array may have different size, let say $e_i \times e_i$ for $1 \leq i \leq p$ where each e_i may be $8, 16$, or 24. We then construct an $e_1 e_2 \cdots e_p$-sized matrix M by the tensor product of these p matrices:

$$M = \bigotimes P_i := P_1 \otimes P_2 \otimes \cdots \otimes P_p.$$

The ciphertext then is $c = M(m \oplus g)$. With this construction we eliminate any possible sparsity of zeros in the encryption matrix M. We note that the key in this case is the entries of the first rows of P_1 to P_p, hence is an array of numbers of size $e_1 + e_2 + \cdots + e_p$ and therefore it is relatively small. The notation $m \oplus g$ means that m is concatenated with g.

13.4   Encryption in Practice

In this section, we consider practical aspects of the cryptosystems we have presented so far in terms of encryption and decryption. In particular, we encrypt the same plaintext $M =$"SBASBA" of length 6 bits with a Hadamard cipher generated by a Hadamard matrix of order 16.

- $C = $ ENCRYPT('SBASBA',16) "Encrypt with H_{16}" $\Rightarrow$
- $C = $ kaia?gcakaia?gca "Identical ciphertext blocks"

It can easily observed that same plaintext blocks results in the same ciphertext blocks and this fact possesses a significant weakness of this cipher when encrypting in ECB mode. However, when we encrypt the same plaintext M with a Hadamard cipher generated by a Hadamard matrix of order $24 = 4 \cdot 6$ we can see that the encryption process does not result in any repetition blocks.

- $C = $ ENCRYPT('SBASBA',24) "Encrypt with H_{24}" $\Rightarrow$
- $C = $ ftaberhzia?wsteinbdarsfa "No repetition blocks"

13.4.1   Electronic Codebook (ECB) Mode

We can now discuss in detail this weakness in the design of the combinatorial design ciphers which in some cases can be eliminated using their iterated versions of product ciphers. As already noted, in cases the plaintext has more than n letters, we repeat the encryption process. This method is also known as the *electronic codebook* mode or ECB in the literature [12, 29, 31, 47]. A disadvantage of this method is that if two plaintext blocks are the same, then the corresponding ciphertext blocks will be identical, and that is visible to the attacker.

The "blow-up" construction can reduce the amount of information that can be retrieved from a potential attacker when using ECB mode by restricting the available choices for combinatorial designs (Hadamard and Williamson Hadamard matrices, orthogonal and Plotkin arrays) A_i, $i = 1, \ldots, k$ to be $A_f \neq A_g$ for $i \leq f, g \leq k$ with $f \neq g$. In general, if we choose the A_i encryption matrices to have $\sum_{i=1}^{k} n_i = n$, where n is the size of the plaintext this weakness is eliminated since the encryption process does not have any repetition blocks (Fig. 13.1).

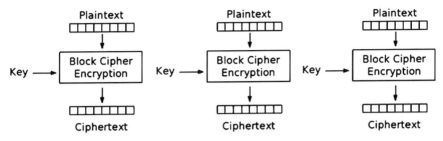

Electronic Codebook (ECB) mode encryption

Fig. 13.1 Illustration of the encryption process in ECB mode

13.4.2 *Diffusion*

Diffusion means that the output bits of the ciphertext should depend on the input bits of the plaintext in a very complex way. In a cipher with good diffusion, if one bit of the plaintext is changed, then the ciphertext should change completely, in an unpredictable or pseudorandom manner. In particular, for a randomly chosen input, if one flips the ith bit, then the probability that the jth output bit will change should be one half, for any i and j. This is termed the strict avalanche criterion, see [51]. More generally, one may require that flipping a fixed set of bits should change each output bit with probability one half. In practice, if one bit of the plaintext is changed, then the ciphertext should change in 2–5 bits in an unpredictable manner. We study the diffusion properties of the Hadamard cipher below.

- $C_1 = \text{ENCRYPT}('1000\ 0001',8) \Rightarrow C_1 = 1100\ 1100$
- $C_2 = \text{ENCRYPT}('0000\ 0001',8) \Rightarrow C_2 = 1000\ 1000$
- $\text{HAMMINGDISTANCE}(C_1,C_2) = 2$

We can easily see that a change in one bit of the original plaintext results in a change of two ciphertext bits using a Hadamard cipher (as this is verified by calculating the Hamming distance of the resulting ciphertext blocks); thus it incorporates the diffusion property.

13.5 Cryptanalysis

The main cryptographic attacks can be classified in the following three categories:

- Brute-force attack
- Plaintext attack
- Ciphertext attack

In this section we demonstrate that our ciphers are robust against brute-force attacks and ciphertext-only attacks, whilst considering some restrictions the corresponding encryption schemes are secure under known-plaintext attacks, chosen-plaintext attacks, and chosen-ciphertext attacks.

13.5.1 Cryptanalysis of Brute-Force Attacks

Definition 13.33 (Brute-Force Attack). A brute-force attack is a method of defeating a cryptographic scheme by trying a large number of possibilities. For most ciphers, a brute-force attack typically means a brute-force search of the key space; that is, testing all possible keys in order to recover the plaintext used to produce a particular ciphertext.

13.5.1.1 Cryptanalysis of Brute-Force Attacks for OA Ciphers

For a given orthogonal array A, with parameters $(n, q, 2, t)$, we can perform $n!$ permutations of rows, $q!$ permutations of columns, and 2^q permutations of the symbols of the columns. Thus we can create $n!q!2^q$ isomorphic representatives for an orthogonal array. We note that the different ones that can result in different ciphertext (see Remark 13.10) are

$$\frac{n!q!2^q}{Aut(A)}$$

where with $Aut(A)$ is the automorphism group of A. It is easy to verify that column permutations and the permutations of the symbols would create create different representatives, in other words for a given $OA(n, q, 2, t)$ there are at least $q!2^q$ different isomorphic orthogonal arrays.

If we denote with $A_{(n,q)}$ the number of non-isomorphic $OA(n, q, 2, t)$ then we can create at least $A_{(n,q)} \cdot q!2^q$ different orthogonal arrays of such parameters. We should mention that for $n > 40$ and $q > 6$ there are no known full lists of non-isomorphic orthogonal arrays. Therefore, it is almost impossible for an adversary to attack on such system using brute force.

Example 13.34. The number of non-isomorphic orthogonal arrays with given parameters $OA(36, 6, 2, t)$ is $A_{(36,6)} = 3352528$, [2]. So we can generate $A_{(36,6)} \cdot 6!2^6 \approx 1.5 \times 10^{12}$ different representatives. The probability of breaking the system via a brute-force attack for this case is less than $P = \frac{1}{1.5 \times 10^{12}} \approx 0.6 \times 10^{-12}$.

Lemma 13.35 (Koukouvinos et al. [24]). *The OA ciphers based on orthogonal arrays with large n, q are secure against brute-force attacks.*

13.5.1.2 Cryptanalysis of Brute-Force Attacks for Hadamard Ciphers

One way for an adversary to break any of the proposed systems using brute-force attack is to generate all possible matrices with elements ± 1, that is 2^{n^2} matrices, having in mind that Hadamard matrices of order n are represented by n^2 bits. However due to the structure of these matrices there exists a more sophisticated method that would be developed next.

13.5.1.3 Cryptanalysis of Brute-Force Attacks for Hadamard Core Ciphers

In order for an adversary to break this system using a brute force attack, he would have to deduce the encryption key $k = A_c$, which is the binary vector $A_c = [a_1, a_2, \ldots, a_p]$ of length p by trying a large number of possibilities.

In our case, an adversary would have to simulate a brute-force search of the key space. Assuming the adversary has knowledge of the encryption protocol he would have to search on p binary variables. Since the encryption key consists of binary variables using enumerative combinatorics, the size of the key space, $K(\mathcal{H}_p)$, is $\mid K(\mathcal{H}_p) \mid = 2^p$; therefore its computational complexity is of exponential growth $O(2^n)$ as $n = p + 1$ increases. Furthermore, the possibility a solution obtained from a brute-force search of the key space to be an encryption key is given by the total number of Hadamard matrices with one circulant core that exists in a specific order divisible by the size of the key space in that order.

For example, if we consider schemes that are using the Hadamard matrices of order $24 = 23 + 1$, the key space consists of 23 binary variables while the total number of Hadamard matrices that exist in that order are 46; therefore we have 46 possible encryption keys. As can be seen in the following table, the probability of breaking the system via a brute-force attack for this case is $P = \frac{46}{2^{23}} \approx 0.00002$, only. It is worthwhile to note that using a key of length only 23 bits, we almost provide total security against brute-force attacks for this scheme.

We summarize in the following table the available Hadamard matrices with one circulant core, denoted by $\mid V(\mathcal{H}_p) \mid$, for orders $n = p + 1$ whereas $\ell = 3, 7, 11, 15, 19, 23$ using the results obtained via exhaustive searches in [21], the cardinality of the key space $\mid K(\mathcal{H}_p) \mid$, and the probability P_{BA} of breaking the cipher via a brute-force attack for each order.

p	Matrix order	$\mid V(\mathcal{H}_p) \mid$	$\mid K(\mathcal{H}_p) \mid = 2^p$	$P_{BA} = \frac{\mid V(\mathcal{H}_p) \mid}{\mid K(\mathcal{H}_p) \mid}$
3	4	3	2^3	$P = \frac{3}{2^3} \approx 0.375$
7	8	14	2^7	$P = \frac{14}{2^7} \approx 0.1$
11	12	22	2^{11}	$P = \frac{22}{2^{11}} \approx 0.01$
15	16	30	2^{15}	$P = \frac{30}{2^{15}} \approx 0.0009$
19	20	38	2^{19}	$P = \frac{38}{2^{19}} \approx 0.00007$
23	24	46	2^{23}	$P = \frac{46}{2^{23}} \approx 0.00002$

As it can be seen from the previous table the sequence of probabilities P_{BA} is strictly decreasing. Based on these computational results we deduce the following remark, when the order n is large enough.

Remark 13.36. The encryption scheme based on Hadamard matrices with one circulant core is secure against brute-force attacks.

Modern cryptographic hardware breakers have the ability to perform a brute-force search for 2^{128} keys. This gives us an estimate of the security needed against brute-force attacks. Clearly, the usage of any Hadamard matrix of order $n > 128$, which can easily be constructed from Theorem 13.15 for large orders, as an encryption matrix justifies our previous claim.

13.5.1.4   Cryptanalysis of Brute-Force Attacks for Hadamard Cores Ciphers

In order for an adversary to break this system using a brute-force attack, he would have to deduce the encryption key $k = A_c \oplus B_c$, which is the concatenation of the binary vectors $A_c = [a_1, a_2, \ldots, a_\ell]$ and $B_c = [b_1, b_2, \ldots, b_\ell]$, of total length 2ℓ by trying a large number of possibilities.

In our case, an adversary would have to simulate a brute-force search of the key space. Assuming the adversary has knowledge of the encryption protocol he would have to search on 2ℓ binary variables. Since the encryption key consists of binary variables using enumerative combinatorics, the size of the key space, $K(\mathcal{H}_\ell)$, is $| K(\mathcal{H}_\ell) | = 2^{2\ell}$; therefore its computational complexity is of exponential growth $O(2^n)$ as $n = 2\ell + 2$ increases. Furthermore, the possibility a solution obtained from a brute-force search of the key space to be an encryption key is given by the total number of Hadamard matrices with two circulant cores that exist in a specific order divisible by the size of the key space in that order.

For example, if we consider schemes that are using the Hadamard matrices of order $28 = 2 \times 13 + 2$, the key space consists of 26 binary variables while the total number of Hadamard matrices that exist in that order are 7,098; therefore we have 7,098 possible encryption keys. As can be seen in the following table, the probability of breaking the system via a brute-force attack for this case is $P = \frac{42 \times 13^2}{2^{26}} \approx 0.0001$, only. It is worthwhile to note that using a key of length only 26 bits we almost provide total security against brute-force attacks for this scheme.

We summarize in the following table the available Hadamard matrices with two circulant cores, denoted by $| V(\mathcal{H}_\ell) |$, for orders $n = 2\ell + 2$ whereas $\ell = 3, \ldots, 25$ using the results obtained via exhaustive searches in [13, 22], the cardinality of the key space $| K(\mathcal{H}_\ell) |$, and the probability P_{BA} of breaking the cipher via a brute-force attack for each order.

As it can be seen from the previous table the sequence of probabilities P_{BA} is strictly decreasing and using (cf. [22, Property 1.]) is upper bounded from 1. In addition, asserting the truth of [22, Conjecture 1.] that for every odd $\ell = 3, \ldots$ there exists a Hadamard matrix of order $2\ell + 2$ with two circulant cores and that the

ℓ	Matrix order	$\lvert V(\mathcal{H}_\ell)\rvert$	$\lvert K(\mathcal{H}_\ell)\rvert = 2^{2\ell}$	$P_{BA} = \dfrac{\lvert V(\mathcal{H}_\ell)\rvert}{\lvert K(\mathcal{H}_\ell)\rvert}$
3	8	$9 = 1 \times 3^2$	2^6	$P = \frac{1 \times 3^2}{2^6} \approx 14 \cdot 10^{-2}$
5	12	$50 = 2 \times 5^2$	2^{10}	$P = \frac{2 \times 5^2}{2^{10}} \approx 4 \cdot 10^{-2}$
7	16	$196 = 4 \times 7^2$	2^{14}	$P = \frac{4 \times 7^2}{2^{14}} \approx 10 \cdot 10^{-3}$
9	20	$972 = 12 \times 9^2$	2^{18}	$P = \frac{12 \times 9^2}{2^{18}} \approx 4 \cdot 10^{-3}$
11	24	$2{,}904 = 24 \times 11^2$	2^{22}	$P = \frac{24 \times 11^2}{2^{22}} \approx 7 \cdot 10^{-4}$
13	28	$7{,}098 = 42 \times 13^2$	2^{26}	$P = \frac{42 \times 13^2}{2^{26}} \approx 10 \cdot 10^{-5}$
15	32	$38{,}700 = 172 \times 15^2$	2^{30}	$P = \frac{172 \times 15^2}{2^{30}} \approx 3 \cdot 10^{-5}$
17	36	$93{,}058 = 322 \times 17^2$	2^{34}	$P = \frac{322 \times 17^2}{2^{34}} \approx 5 \cdot 10^{-6}$
19	40	$161{,}728 = 448 \times 19^2$	2^{38}	$P = \frac{488 \times 19^2}{2^{38}} \approx 5 \cdot 10^{-7}$
21	44	$433{,}944 = 984 \times 21^2$	2^{42}	$P = \frac{984 \times 21^2}{2^{42}} \approx 10 \cdot 10^{-8}$
23	48	$1{,}235{,}744 = 2336 \times 23^2$	2^{46}	$P = \frac{2336 \times 23^2}{2^{46}} \approx 2 \cdot 10^{-8}$
25	52	$2{,}075{,}000 = 3320 \times 25^2$	2^{50}	$P = \frac{3320 \times 25^2}{2^{50}} \approx 2 \cdot 10^{-9}$

sequence of $\lvert V(\mathcal{H}_\ell)\rvert$ will continue to increase we can conclude that the limit of the sequence of probabilities $\lim_{\ell \to \infty} P_{BA} = \lim_{\ell \to \infty} \frac{\lvert V(\mathcal{H}_\ell)\rvert}{\lvert K(\mathcal{H}_\ell)\rvert}$ converges to zero. Note that Conjecture 1 of [22] would settle the general Hadamard conjecture. In particular, we quote the following lemma.

Lemma 13.37 (Koukouvinos and Simos [26]). *Assume the following two conditions hold,*

(i) *There exists a Hadamard matrix of order $2\ell + 2$ with two circulant cores for every odd $\ell = 3, \ldots$*
(ii) *The sequence of $\lvert V(\mathcal{H}_\ell)\rvert$ is increasing for every odd $\ell = 3, \ldots$*

Then, the encryption scheme based on Hadamard matrices with two circulant cores is secure against brute-force attacks.

13.5.1.5 Simulation of Brute-Force Attacks for Plotkin Ciphers

To carry a brute-force attack on the Plotkin cipher we carried the following steps for each simulation.

1. We used a sample plaintext of 384 characters and a random noise vector of the same length.
2. We considered the entries of $A, B, \ldots, H$ as binary variables.
3. We decoded the ciphertext using every key combination of key entry and key entry value equal to ± 1.

From the experimental results we received from the first encryption scheme we obtained the following information:

1. For the Plotkin arrays $OD(8t; t, t, t, t, t, t, t, t)$ for $t = 8, 16, 24$ a brute-force attack resulted in a thorough defeat of the cipher. We mention though that the computational time grows in a nonlinear way.

Table 13.1 Experimental results received from a brute-force attack on the Plotkin cipher

Design	Key search space	Elapsed CPU time (h)
$OD(8; 1, 1, 1, 1, 1, 1, 1, 1)$	2^8	4
$OD(16; 2, 2, 2, 2, 2, 2, 2, 2)$	2^8	12
$OD(24; 3, 3, 3, 3, 3, 3, 3, 3)$	2^8	34

Algorithm 4 AnalyzerScheme Function

function ANALYZERSCHEME(Receives the output from the HackerFunction and calculates the frequency of occurrence of every ASCII symbol)

    Step. 1 For each line of text, count number of appearances of each ASCII value.
    Step. 2 Output information to text file.

end function

---

2. Since this scheme is not robust against brute attacks we have a complete violation to one of the design properties we set in the introduction for this encryption scheme.

Table 13.1 presents the computational results for the simulations we conducted. For each orthogonal design we give the size of the key search space and the elapsed CPU time needed for a brute-force attack to break the system.

13.5.1.6 Simulation of Brute-Force Attacks for Kronecker Plotkin Ciphers

To carry a brute-force attack on the Kronecker Plotkin ciphers we carried the following steps for each simulation.

1. We used a sample plaintext of 23 characters.
2. We encoded the plaintext using the second scheme by approximating the entry size for the Plotkin arrays and approximate size of the noise vector g.
3. We used the Plotkin arrays of order 8 to compute the encryption matrix M.
4. We decoded the ciphertext using every key combination of key entry and key entry value equal to ± 1.
5. We converted the decoded ciphertext found in the previous step to ASCII values.
6. We counted the frequency of each value that appears in the resulting combinations.

We provide also the following cryptographic algorithms if someone wants to implement the previous simulation procedure in an efficient manner.

Table 13.2 Experimental results received from a brute-force attack on the Kronecker Plotkin cipher

		ASCII values occurrences$\times 10^5$				
key size	noise size	$0-25$	$26-50$	$51-75$	$76-100$	$101-127$
10–14	128	25	5	5	7	8
10–14	1024	10	12	8	6	14
30–34	128	120	30	40	30	50
30–34	1024	65	90	45	50	40
50–54	128	310	50	70	30	40
50–54	1024	110	100	90	80	120

Algorithm 5 HackerScheme Function

---

function HACKERSCHEME(Simulation of a brute force attack method to a ciphertext)

Step 1. Input min, max, and range of key guesses.

Step 2. Input ciphertext.

Step 3. Exhaustive key search with respect to Step 1.

For all possible values of the variables of the orthogonal designs chosen for encryption perform the following steps.

Step 3a. Generate the matrices using as entries the possible values from previous step.

Step 3b. Compute the tensor product of the matrices created in previous step.

Step 3c. Calculate possible text messages.

Step 3d. Output text to file for later examination.

end function

From the experimental results we received for the Kronecker Plotkin cipher we obtained the following information:

1. A brute-force attack is not a feasible way of defeating the cipher.
2. One advantage of the one time pad is that a brute-force attack results in all possible plaintext messages, forcing an adversary to choose which was the original message. We wanted to determine if this was also true for our cipher. The computational results indicate that the answer is no.
3. Finally we wanted to determine if the size of the entries of the noise vector g played a significant role in the decryption process. The computations showed that the answer is yes.
4. All design goals are fulfilled for this encryption scheme.

The above Table 13.2 presents the computational results for the simulations we conducted. For each simulated brute-force attack we give the number of occurrences of the ASCII values in their corresponding range and the approximate key and noise vector sizes. The table shows that most of the characters that appear in the simulated brute-force attack are those that have been encoded using the sample plaintext.

13.5.2 Cryptanalysis of Known-Plaintext Attacks for Hadamard and Plotkin Ciphers

Definition 13.38 (Known-plaintext Attack). A known-plaintext attack is one where the adversary has a quantity of plaintext and corresponding ciphertext. This type of attack is typically only marginally more difficult to mount.

Supposing a $n \times n$ matrix A is used for encryption, as described previously. In order to recover the matrix $A = H_n$ of a Hadamard cipher or $A = P$ for a Plotkin cipher without knowing the private key, we will need n $\overline{m}^i$'s, where with $\overline{m}^i = (m_1^i, m_2^i, \ldots, m_n^i)$, $i = 1, \ldots, n$ we denote the vector consisting of n letters of the message that have been converted to its numerical values, and n $\overline{c}^i$'s, where each $\overline{c}^i = (c_1^i, c_2^i, \ldots, c_n^i)$ is the encryption of $\overline{m}^i$. We can retrieve the ith column of A, $A(i) = (a_{1,i}, a_{2,i}, \ldots, a_{n,i})$, by solving the following n-linear systems, for $i = 1, \ldots, n$:

$$m_1^1 a_{1,i} + m_2^1 a_{2,i} + \cdots + m_n^1 a_{n,i} = c_i^1$$
$$m_1^2 a_{1,i} + m_2^2 a_{2,i} + \cdots + m_n^2 a_{n,i} = c_i^2$$
$$\vdots \quad \vdots$$
$$m_1^n a_{1,i} + m_2^n a_{2,i} + \cdots + m_n^n a_{n,i} = c_i^n$$

or equivalently we denote the previous system

$$MA(i) = C(i),$$

where $C(i) = (c_i^1, c_i^2, \ldots, c_i^n)$.

Proposition 13.39 (Koukouvinos and Simos [25, 26]). *Hadamard and Plotkin ciphers are secure against known-plaintext attacks under the assumption that the adversary has knowledge of less than n messages of length n of the plaintext and the corresponding ciphertext.*

13.5.3 Cryptanalysis of Chosen-Plaintext Attacks for Hadamard and Plotkin Ciphers

Definition 13.40 (Chosen-Plaintext Attack). A chosen-plaintext attack is one where the adversary chooses plaintext and is then given corresponding ciphertext. Subsequently, the adversary uses any information deduced in order to recover plaintext corresponding to previously unseen ciphertext.

In this type of attack the extra advantage of the adversary having knowledge of the encryption mechanism does not reveal any further information with respect to a known-plaintext attack since the adversary in order to compromise the system still has to solve n linear systems,

$$MA(i) = C(i)$$

for $i = 1, \ldots, n$ as described in Sect. 13.5.2.

Remark 13.41. The adversary should take under account that the matrix M of the chosen plaintext must not be singular. This note restricts the choice of the available plaintexts for an adversary since $\overline{m}^i \neq \lambda \overline{m}^j$, in other words the vectors $\overline{m}^i$ must be linear independent.

Proposition 13.42 (Koukouvinos and Simos [25, 26]). *Hadamard and Plotkin ciphers are secure against chosen-plaintext attacks, since the schemes are secure against known-plaintext attacks.*

13.5.4 Cryptanalysis of Chosen-Ciphertext Attacks for Hadamard and Plotkin Ciphers

Definition 13.43 (Chosen-Ciphertext Attack). A chosen-ciphertext attack is one where the adversary selects the ciphertext and is then given the corresponding plaintext. One way to mount such an attack is for the adversary to gain access to the equipment used for decryption (but not the decryption key, which may be securely embedded in the equipment). The objective is then to be able, without access to such equipment, to deduce the plaintext from (different) ciphertext.

Similar, in this type of attack the extra advantage of the adversary having knowledge of the encryption mechanism, does not reveal any further information with respect to a known-plaintext attack since the adversary in order to compromise the system still has to solve n linear systems,

$$MA(i) = C(i)$$

for $i = 1, \ldots, n$ as described in Sect. 13.5.2.

Proposition 13.44 (Koukouvinos and Simos [25, 26]). *Hadamard and Plotkin ciphers are secure against chosen-ciphertext attacks, since the schemes are secure against known-plaintext attacks.*

13.5.5 Cryptanalysis of Known-Plainext, Chosen-Plaintext and Ciphertext Attacks for Kronecker Hadamard and Plotkin Ciphers

An intriguing question is if the security provided by the Hadamard and Plotkin ciphers is enough for standard applications (i.e., banking transactions) in practice. Clearly, the security is a function of the value n of the plaintext's length.

For example, with a plaintext of $n = 64$ bits an attacker which can deduce $64 = 2^6$ messages of the same length can break the ciphers and of course this is totally impractical!

The solution to this problem is to use the Kronecker Hadamard and Plotkin ciphers. For example, using 16 rounds of encryption, i.e., the Kronecker product of 16 Hadamard matrices or Plotkin arrays of order 16 the size of the encryption matrix is $2^{4^{16}} = 2^{64}$, while the key size is $16 \times 15 = 240$ bits. Therefore, using a key of 240 bits we provide security for 2^{64} known and chosen-plaintexts and ciphertexts. We compare now this result with the security of a widely known modern block cipher, i.e., DES.

1. To break the full 16-rounds of DES Bilham and Shamir showed that differential cryptanalysis requires 2^{47} chosen plaintexts (see [3, 4]).
2. Linear cryptanalysis discovered by Matsui needs 2^{43} known plaintexts to achieve similar results (see [30]).

13.5.6   Cryptanalysis of Ciphertext-Only Attacks for Hadamard and Plotkin Ciphers

Definition 13.45 (Ciphertext-only Attack).   A ciphertext-only attack is one where the adversary (or cryptanalyst) tries to deduce the decryption key or plaintext by only observing ciphertext. Any encryption scheme vulnerable to this type of attack is considered to be completely insecure.

Two letters of the original message, m, correspond to different values of the ciphertext, $\bar{c}$. Analyzing the worst-case scenario for this type of attack, we suppose that all letters of the plaintext are the same. Then in the corresponding ciphertext all their numerical values are all different. Therefore an adversary cannot observe any further information regarding the encryption key or the plaintext, since any value of the encrypted message is a function of n values of the plaintext and one column of the encryption matrix A. Hence, two or more same values of the encrypted message does not represent the same letter in the plaintext. We note that, as n increases it is more difficult for an adversary to retrieve the encryption key or the plaintext by simple observation.

Proposition 13.46 (Koukouvinos and Simos [25, 26]).   *Hadamard and Plotkin ciphers are secure against ciphertext-only attacks.*

13.5.7   Security Comparison for Combinatorial Design Ciphers

As we have seen, for example, a chosen-plaintext attack can break the Hadamard and Plotkin ciphers. However, with a key length ≥ 128 bits we provide security

Table 13.3 Security comparison for combinatorial design-based ciphers

Cipher family	Block size	Key length	Key size
OA cipher	n bits	qn bits	$\mathcal{O}(n^2)$
Hadamard cipher	n bits	n^2 bits	$\mathcal{O}(n^2)$
Hadamard core cipher	n bits	$(n-1)$ bits	$\mathcal{O}(n)$
Hadamard cores cipher	n bits	$(n-2)$ bits	$\mathcal{O}(n)$
Williamson cipher	n bits	n bits	$\mathcal{O}(n)$
Plotkin cipher	n bits	n^2 bits	$\mathcal{O}(n^2)$

against brute-force attacks. For 3DES (Triple DES or three rounds of encryption of DES) there exists a meet-in-the-middle attack that provides security only for 112 bits, when using a key of 168 bits (three 56 bit DES keys), for more details see [12]. While Blowfish uses a variable-key size of length up to 448 bits [39].

We conclude this survey paper by giving a comparison of the security provided by the presented combinatorial design ciphers in terms of key length and size under the assumption that all ciphers operate with an encryption matrix of order n (thus the block size of the plaintext is of length n due to ECB mode of encryption) (Table 13.3).

Asides of the iterated combinatorial block ciphers, whose advantages were illustrated in Sect. 13.5.5, from the previous comparison is clear that the Hadamard cores cipher is the most efficient when compared with the whole class of the combinatorial design-based ciphers we presented in this paper.

References

1. Angelopoulos, P., Evangelaras, H., Koukouvinos, C., Lappas, E.: An effective step-down algorithm for the construction and the identification of nonisomorphic orthogonal arrays. Metrika. **66**, 139–149 (2007)
2. Angelopoulos, P., Koukouvinos, C., Lappas, E.: On construction, classification and evaluation of certain two level nonisomorphic orthogonal arrays. Int. J. Appl. Math. Stat. **15**, 63–72 (2009)
3. Biham, E., Shamir, A.: Differential Cryptanalysis of DES-like Cryptosystems. Advances in Cryptology CRYPTO '90, pp. 2–21. Springer-Verlag (1990)
4. Biham, E., Shamir, A.: Differential Cryptanalysis of the Full 16-Round DES, CS 708. In: Proceedings of CRYPTO '92. Lecture Notes in Computer Science, Vol. 740 (1991)
5. Boyd, C., Mathuria, A.: Protocols for Authentication and Key Establishment. Information Security and Cryptography Series. Springer-Verlag, Heidelberg (2003)
6. Brown, L., Pieprzyk, J., Seberry, J.: LOKI - a cryptographic primitive for authentication and secrecy applications. In: Seberry, J., Pieprzyk, J. (ed.) Advances in Cryptology - Auscrypt'90, LNCS 453, pp. 229–236. Springer-Verlag (1990)
7. Bulutoglu, D.A., Margot, F.: Classification of orthogonal arrays by integer programming. J. Statist. Plann. Inference. **138**, 654–666 (2008)
8. Colbourn, C.J., Dinitz, J.H., Stinson, D.R.: Applications of combinatorial designs to communications, cryptography, and networking. In: Lamb, J.D., Preece, D.A. (eds.) Surveys in Combinatorics, pp. 37–100. Cambridge University Press, Cambridge (1999)
9. Cormen, T.H., Leiserson, C.H., Rivest, R.L., Stein, C.: Introduction to Algorithms. MIT Press (2003)

10. Craigen, R.: Hadamard matrices and designs. In: Colbourn, C.J., Dinitz, J.H. (eds.) The CRC Handbook of Combinatorial Designs, pp. 370–377. Crc Press, Boca Raton, FL (1996)
11. Evangelaras, H., Koukouvinos, C., Lappas, E.: Further contributions to nonisomorphic two level orthogonal arrays. J. Statist. Plann. Inference. **137**, 2080–2086 (2007)
12. Ferguson, N., Schneier, B.: Practical Cryptography. Wiley Publishing, Inc. (2003)
13. Fletcher, R.J., Gysin, M., Seberry, J.: Application of the discrete Fourier transform to the search for generalised Legendre pairs and Hadamard matrices. Australas. J. Combin. **23**, 75–86 (2001)
14. Georgiou, S., Koukouvinos, C.: On generalized Legendre pairs and multipliers of the corresponding supplementary difference sets. Utilitas Math. **61**, 47–63 (2002)
15. Georgiou, S., Koukouvinos, C., Seberry, J.: Hadamard matrices, orthogonal designs and construction algorithms, Chapter 7. In: Wallis, W.D. (ed.) Designs 2002: Further Computational and Constructive Design Theory, pp. 133–205. Kluwer Academic Publishers, Norwell, Massachusetts (2003)
16. Geramita, A.V., Seberry, J.: Orthogonal Designs: Quadratic Forms and Hadamard Matrices. Marcel Dekker, New York-Basel (1979)
17. Gysin, M., Seberry, J.: An experimental search and new combinatorial designs via a generalization of cyclotomy. J. Combin. Math. Combin. Comput. **27**, 143–160 (1998)
18. Hadamard, J.: Resolution d'une question relative aux determinants. Bull. des. Sci. Math. **17**, 240–246 (1893)
19. Hall, M. Jr.: A survey of difference sets. Proc. Amer. Math. Soc. **7**, 975–986 (1956)
20. Hedayat, A.S., Sloane, N.J.A., Stufken, J.: Orthogonal Arrays: Theory and Applications. Springer-Verlag, New York (1999)
21. Kotsireas, I.S., Koukouvinos, C., Seberry, J.: Hadamard ideals and Hadamard matrices with circulant core. J. Combin. Math. Combin. Comput. **57**, 47–63 (2006)
22. Kotsireas, I.S., Koukouvinos, C., Seberry, J.: Hadamard ideals and Hadamard matrices with two circulant cores. European J. Combin. **27**, 658–668 (2006)
23. Koukouvinos, C.: Williamson matrices. [Online]. Available: http://www.math.ntua.gr/~ckoukouv/designs.htm
24. Koukouvinos, C., Lappas, E., Simos, D.E.: Encryption schemes using orthogonal arrays. J. Discrete Math. Sci. Cryptogr. **12**, 615–628 (2009)
25. Koukouvinos, C., Simos, D.E.: Encryption schemes using plotkin arrays. Appl. Math. Inf. Sci. **5**, 500–510 (2011)
26. Koukouvinos, C., Simos, D.E.: Encryption schemes based on hadamard matrices with circulant cores. submitted for publication.
27. van Lint, J.H., Wilson, R.M.: A Course in Combinatorics. Cambridge University Press, Cambridge (1992)
28. Luby, M.: Pseudorandomness and Cryptographic Applications. Princeton Academic Press, Princeton (1996)
29. Mao, W.: Modern Cryptography: Theory and Practice. Prentice Hall (2004)
30. Matsui, M.: Linear cryptanalysis method for DES cipher. In Workshop on the theory and application of cryptographic techniques on Advances in cryptology (EUROCRYPT '93), Tor Helleseth (Ed.). Springer-Verlag New York, Inc., Secaucus, NJ, USA, 386–397 (1994)
31. Menezes, A., van Oorschot, P., Vanstone, S.: Handbook of Applied Cryptography. CRC Press (1997)
32. Orrick, W.: Switching operations for Hadamard matrices. SIAM J. Discr. Math. **22**, 31–50 (2008)
33. Paley, R.E.A.C.: On orthogonal matrices. J. Math. Phys. **12**, 311–320 (1933)
34. Plotkin, M.: Decomposition of Hadamard matrices. J. Combin. Theory, Ser. A. **13**, 127–130 (1972)
35. Rao, C.R.: Factorial experiments derivable from combinatorial arrangements of arrays. J. Royal Stat. Society (Suppl.). **9**, 128–139 (1947)
36. Rao, C.R.: On a class of arrangements. Proc. Edinburgh Math. Society. **8**, 119–125 (1949)
37. Sarvate, D.G., Seberry, J.: Encryption methods based on combinatorial designs. Ars Combinatoria. **21**-A, 237–246

38. Schneier, B.: Description of a New Variable-Length Key, 64-bit Block Cipher (Blowfish). Fast Software Encryption 1993: 191–204
39. Schneier, B.: Description of a New Variable-Length Key, 64-bit Block Cipher (Blowfish). In Fast Software Encryption, Cambridge Security Workshop, Ross J. Anderson (Ed.). Springer-Verlag, London, UK, 191–204 (1993)
40. Schroeder, M.R.: Number Theory in Science and Communication. Springer–Verlag, New York (1984)
41. Seberry, J., Craigen, R.: Orthogonal designs. In: Colbourn, C.J., Dinitz, J.H. (eds.) CRC Handbook of Combinatorial Designs, pp. 400–406. CRC Press, Boca Raton (1996)
42. Seberry, J., Yamada, M.: Hadamard matrices, sequences and block designs. In: Dinitz, J.H., Stinson, D.R. (eds.) Contemporary Design Theory: A Collection of Surveys, pp. 431–560. J. Wiley and Sons, New York (1992)
43. Shimizu, A., Miyaguchi, S.: Fast data encipherment algorithm FEAL, Advances in Cryptology Eurocrypt '87, pp. 267–280. Springer-Verlag (1988)
44. Singer, J.: A theorem in finite projective geometry and some applications to number theory. Trans. Amer. Math. Soc. **43**, 377–385 (1938)
45. Stanton, R.G., Sprott, D.A.: A family of difference sets. Can. J. Math. **10**, 73–77 (1958)
46. Stallings, W.: Cryptography and Network Security: Principles and Practices, 3rd Edn. Prentice Hall (2003)
47. Stinson, D.R.: Cryptography: Theory and Practice, 3rd Edn. CRC Press (2005)
48. Stufken, J., Tang, B.: Complete enumeration of two-level orthogonal arrays of strength d with $d + 2$ constraints. Ann. Statist. **35**, 793–814 (2007)
49. Sylvester, J.J.: Thoughts on inverse orthogonal matrices, simultaneous sign-successions, and tessellated pavements in two or more colors, with applications to Newtons rule, ornamental tile-work, and the theory of numbers. Phil. Mag. **34**, 461–475 (1867)
50. Turyn, R.J.: An infinite class of Williamson matrices. J. Combin. Theory Ser. A. **12**, 319–321 (1972)
51. Webster, A.F., Tavares, E.S.: On the design of S-boxes, Advances in Cryptology - Crypto '85. Lecture Notes in Computer Science, Vol. 219, pp. 523–534. Springer-Verlag Inc., New York, NY (1985)
52. Williamson, J.: Hadamard's determinant theorem and the sum of four squares. Duke Math. J. **11**, 65–81 (1944)
53. Whiteman, A.L.: An infinite family of Hadamard matrices of Williamson type. J. Combin. Theory Ser. A. **14**, 334–340 (1973)
54. Whiteman, A.L.: A family of difference sets. Illinois J. Math. **6**, 107–121 (1962)

Chapter 14
On the Weak Convergence of an Empirical Estimator of the Discrete-Time Semi-Markov Kernel

Stylianos Georgiadis and Nikolaos Limnios

Abstract In this chapter, we consider the discrete-time semi-Markov processes with finite state space and the empirical estimator of the semi-Markov kernel. The basic definitions concerning the semi-Markov processes are presented. We study the weak convergence of the empirical estimator of the discrete-time semi-Markov kernel. Next, we present the corresponding weak convergence theorem for the empirical estimator of some related measures. The proofs of our results are based on semimartingales.

Keywords Discrete-time semi-Markov kernel • Empirical estimator • Weak convergence • Invariance principle • Semimartingales

AMS Subject Classification: 60F05, 60F17, 60K15, 62G05

14.1 Introduction

Semi-Markov processes constitute a generalization of Markov jump processes and renewal processes. For a Markov process, the sojourn time in each state is exponentially distributed (in discrete time, geometrically distributed), while for the semi-Markov case, the sojourn-time distribution can be any distribution on $\mathbb{R}_+$ (in discrete time, $\mathbb{N}$). In recent literature, semi-Markov models have achieved significant importance in probabilistic and statistical modeling especially the ones

S. Georgiadis (✉) • N. Limnios
Laboratoire de Mathématiques Appliquées, Université de Technologie de Compiègne,
Centre de Recherches de Royallieu, BP 20529, 60205, Compiegne Cedex, France
e-mail: stylianos.georgiadis@utc.fr; Nikolaos.Limnios@utc.fr

N.J. Daras (ed.), *Applications of Mathematics and Informatics in Military Science*, Springer Optimization and Its Applications 71, DOI 10.1007/978-1-4614-4109-0_14, © Springer Science+Business Media New York 2012

with a finite space state. Possible application fields are survival analysis, reliability theory, DNA analysis, statistical seismology, etc. The random evolution of a semi-Markov process is completely characterized by a semi-Markov kernel. In addition, all relative measures in application topics can be expressed as functionals of the semi-Markov kernel. Consequently, the estimation of the semi-Markov kernel has been of foremost importance in the problem of statistical inference for semi-Markov processes.

The basic theory of semi-Markov processes was introduced by Pyke [13, 14]. Nonparametric empirical estimation for semi-Markov kernels has been presented in several works. In finite case, Moore and Pyke [10] studied empirical and maximum likelihood estimators for semi-Markov kernel while Gill [5] gave an estimator using counting processes. Greenwood and Wefelmeyer [6] studied efficiency of empirical estimators for linear functionals in the case of general state space. Ouhbi and Limnios [11, 12] studied empirical estimators for nonlinear functionals of semi-Markov kernel (including Markov renewal matrices and reliability functionals) and their asymptotic properties. Limnios [8] studied the invariance principle for the empirical estimator of semi-Markov kernel and Georgiadis and Limnios [4] extend these results in a multidimensional form. Limnios and Oprişan [9] present a total study of semi-Markov processes on continuous time and their applications in reliability.

In this work, we consider the discrete-time semi-Markov processes. So on, we will use the term chain for a discrete-time process. A thorough presentation on the semi-Markov chains is given toward applications by Barbu and Limnios [2].

Generally, the manipulation of semi-Markov chains seems more convenient than the continuous ones, especially for applications. In discrete time, the Markov renewal function is expressed as a finite series of semi-Markov kernel convolution product, instead of a finite series in the continuous case. Furthermore, a semi-Markov chain makes only a finite number of transitions in a finite time interval. As a result, discrete-time semi-Markov models have more accurate computations and numerical results. Moreover, the semi-Markov chain can be a good support for the numerical calculus of the continuous-time semi-Markov processes, after their discretization. Semi-Markov chains are less studied but, recently, there is a growing literature concerning their inference problems. Barbu et al. [3] introduce the discrete-time SMP, propose a computation procedure for solving the corresponding Markov renewal equation, and, then, compute some reliability measurements. Barbu and Limnios [1] consider a discrete-time finite state space semi-Markov model, introduce an empirical estimator of the DTSMK and other measurements, and study the strong consistency and asymptotic normality for these estimators.

The weak convergence of the discrete-time semi-Markov kernel presented in this work consists of the discrete-time analogous of the results presented in [4]. The present chapter is organized as follows. In Sect. 14.2, the necessary mathematical background of semi-Markov chains and semimartingales is introduced. Next, the main results on the weak convergence of the discrete-time semi-Markov kernel are given. Finally, in Sect. 14.4, we examine the weak convergence of some kernels and functions of the semi-Markov kernel.

14.2 Preliminaries

In this section, we give all the necessary preliminaries concerning the semi-Markov chains and the semimartingales. From now on we will use the following notation: $\mathbb{N}^* = \mathbb{N} \setminus \{0\}$, and $\mathbb{R}_+^* = (0, \infty)$.

14.2.1 Semi-Markov Chains

Consider a finite set E and an E-valued stochastic chain $(Z_k)_{k \in \mathbb{N}}$. Let $(J_n)_{n \in \mathbb{N}}$ be the successive visited states of (Z_k) with state space E and $(S_n)_{n \in \mathbb{N}}$ are the jump times of (Z_k) with values in $\mathbb{N}$ and $0 = S_0 \leq S_1 \leq \cdots \leq S_n \leq S_{n+1} \leq \cdots$. Also, let us denote $X_n := S_n - S_{n-1}, n \in \mathbb{N}^*$, as the sojourn times in these states with values in $\mathbb{N}$.

Definition 14.1. The stochastic process $(J_n, S_n)_{n \in \mathbb{N}}$, with state space E, is said to be a Markov renewal chain (MRC), if, for all $j \in E, k \in \mathbb{N}$ and $n \in \mathbb{N}$, it satisfies a.s. the following equality:

$$\mathbb{P}(J_{n+1} = j, X_{n+1} = k | J_0, \ldots, J_n; S_1, \ldots, S_n) = \mathbb{P}(J_{n+1} = j, X_{n+1} = k | J_n), \quad n \in \mathbb{N}.$$

In this case, (Z_k) is called a semi-Markov chain (SMC).

Actually, (Z_k) gives the state of the process at time k. We assume that the MRC (J_n, S_n) is time homogeneous, i.e., the above probability is independent of n and S_n. The process (J_n) is a Markov chain with state space E and transition kernel $p := (p_{ij}; i, j \in E)$, where

$$p_{ij} := \mathbb{P}(J_{n+1} = j | J_n = i), \tag{14.1}$$

called the embedded Markov chain (EMC) of (Z_k). The SMC (Z_k) is associated with the MRC (J_n, S_n) by

$$Z_k = J_n, \quad S_n \leq k < S_{n+1},$$
$$J_n = Z_{S_n}, \quad n \in \mathbb{N}.$$

We denote by $N(k)$, $k \in \mathbb{N}$, the process which counts the number of jumps of (Z_k) in the interval $[1,k]$, defined by $N(k) := \max\{n \geq 0 : S_n \leq k\}$. The process (Z_k) then can be written as

$$Z_k := J_{N(k)}, \quad k \in \mathbb{N}.$$

Let $N_i(k)$ be the number of visits of (Z_k) to state $i \in E$ up to time k, and $N_{ij}(k)$ the number of direct jumps of (Z_k) from state i to state j up to time k. To be specific,

$$N_i(k) := \sum_{m=1}^{N(k)} \mathbf{1}_{\{J_{m-1}=i\}} \quad \text{and} \quad N_{ij}(k) := \sum_{m=1}^{N(k)} \mathbf{1}_{\{J_{m-1}=i,J_m=j\}},$$

where $\mathbf{1}_A$ is the indicator function of the set A. Considering the renewal process $(S_n^i)_{n \in \mathbb{N}}$ of successive times of visits to state i, $N_i(k)$ is the counting process of renewals.

Definition 14.2. The transition kernel $q(k) := (q_{ij}(k); i, j \in E)$, $k \in \mathbb{N}$, is called the discrete-time semi-Markov kernel (DTSMK) of the SMC (Z_k) and is defined by

$$q_{ij}(k) := \mathbb{P}(J_{n+1} = j, X_{n+1} = k | J_n = i), \tag{14.2}$$

$$p_{ij} := Q_{ij}(\infty) := \lim_{k \to \infty} Q_{ij}(k),$$

We give the definition of some quantities related to the DTSMK.

Definition 14.3. For all $i, j \in E$ and $k \in \mathbb{N}$, the entries of the transition kernel p in terms of the DTSMK are written as

1. $Q(k) = (Q_{ij}(k); i, j \in E)$, the cumulative DTSMK:

$$Q_{ij}(k) := \mathbb{P}(J_{n+1} = j, X_{n+1} \leq k | J_n = i) = \sum_{l=0}^{k} q_{ij}(l), \tag{14.3}$$

2. $f(k) := (f_{ij}(k); i, j \in E)$, the conditional distribution function of the sojourn time in state i, given that the next visited state is j, $j \neq i$:

$$f_{ij}(k) := \mathbb{P}(X_{n+1} = k | J_n = i, J_{n+1} = j) = \begin{cases} \frac{q_{ij}(k)}{p_{ij}}, & \text{if } p_{ij} \neq 0, \\ \mathbf{1}_{\{k=\infty\}}, & \text{if } p_{ij} = 0, \end{cases} \tag{14.4}$$

3. $F(k) := (F_{ij}(k); i, j \in E)$, the conditional cumulative distribution function of the sojourn time in state i, given that the next visited state is j, $j \neq i$:

$$F_{ij}(k) := \mathbb{P}(X_{n+1} \leq k | J_n = i, J_{n+1} = j) = \begin{cases} \frac{Q_{ij}(k)}{p_{ij}}, & \text{if } p_{ij} \neq 0, \\ \mathbf{1}_{\{k=\infty\}}, & \text{if } p_{ij} = 0, \end{cases} \tag{14.5}$$

4. $h(k) := (h_i(k); i \in E)$, the sojourn time distribution function in state i:

$$h_i(k) := \mathbb{P}(X_{n+1} = k | J_n = i) = \sum_{j \in E} q_{ij}(k), \qquad (14.6)$$

5. $H(k) := (H_i(k); i \in E)$, the sojourn time cumulative distribution function in state i:

$$H_i(k) := \mathbb{P}(X_{n+1} \leq k | J_n = i) = \sum_{l=0}^{k} h_i(l) = \sum_{l=0}^{k} \sum_{j \in E} q_{ij}(l), \qquad (14.7)$$

Let us denote by μ_{ii} the mean recurrence times of (S_n^i), i.e., $\mu_{ii} := \mathbb{E}[S_2^i - S_1^i]$ and by $\pi = (\pi_i; i \in E)$ and $v = (v_i; i \in E)$, the stationary distribution of the SMC (Z_k) and the EMC (J_n), respectively. Let us, also, define the mean sojourn time of (Z_k) as $\bar{m} := \sum_{i \in E} v_i m_i$ with m_i to be the mean sojourn time of (Z_k) in state $i \in E$, i.e., $m_i := \sum_{n \in \mathbb{N}} [1 - H_i(n)]$. For an arbitrary state $i \in E$, the following properties hold:

$$v_i = \frac{\bar{m}}{\mu_{ii}} \quad \text{and} \quad \pi_i = \frac{m_i}{\mu_{ii}}.$$

For the whole article, we assume that the EMC (J_n) is irreducible and $\bar{m} < \infty$.

A useful lemma in the proof of our results is the following.

Lemma 14.4 [8]. *If the EMC (J_n) is irreducible with finite stationary distribution v and $\bar{m} < \infty$, then, for any $i, j \in E$, we have*

1. $\frac{N_i(k)}{k} \xrightarrow{a.s.} \frac{1}{\mu_{ii}}$
2. $\frac{N(k)}{k} \xrightarrow{a.s.} \frac{1}{\bar{m}}$
3. $\frac{N_{ij}(k)}{k} \xrightarrow{a.s.} \frac{p_{ij}}{\mu_{ii}}$

as $k \to \infty$.

Definition 14.5 [2]. Let $A(k) := (A_{ij}(k); i, j \in E)$, $B(k) := (B_{ij}(k); i, j \in E)$ be two matrix-valued functions. The discrete-time matrix convolution product $A * B(k)$ is the matrix-valued $C(k) := (C_{ij}(k); i, j \in E)$ defined by

$$C_{ij}(k) := \sum_{r \in E} \sum_{l=0}^{k} A_{ir}(k - l) B_{rj}(l), \quad k \in \mathbb{N},$$

or, in matrix form,

$$C(k) := \sum_{l=0}^{k} A(k - l) B(l), \quad k \in \mathbb{N}.$$

The discrete-time n-fold convolution $q^{(n)}(k)$, $n, k \in \mathbb{N}$, of $q(k)$ by itself can be defined recursively by

$$q_{ij}^{(0)}(k) := \begin{cases} 1, & \text{if } i = j \text{ and } k = 0, \\ 0, & \text{elsewhere,} \end{cases}$$

$$q_{ij}^{(1)}(k) := q_{ij}(k),$$

$$q_{ij}^{(n)}(k) := \sum_{r \in E} \sum_{l=0}^{k} q_{ir}^{(l)}(k) q_{rj}^{(n-1)}(k-l), \quad n \geq 2,$$

or, in matrix form,

$$q^{(0)}(k) := \begin{cases} I, & \text{if } k = 0, \\ 0, & \text{elsewhere,} \end{cases}$$

$$q^{(1)}(k) := q(k),$$

$$q^{(n)}(k) := q * q^{(n-1)}(k), \quad n \geq 2.$$

For any $i, j \in E$ and any $n, k \in \mathbb{N}$, the n-fold convolution $q_{ij}^{(n)}(k)$ can be written as

$$q_{ij}^{(n)}(k) := \mathbb{P}(J_n = j, S_n = k | J_0 = i),$$

Definition 14.6. The Markov renewal function $\psi(k) := (\psi_{ij}(k); i, j \in E)$, is defined by

$$\psi_{ij}(k) := \sum_{n=0}^{k} q_{ij}^{(n)}(k), \quad k \in \mathbb{N},$$

Remark 14.7. The fact that the Markov renewal function can be expressed as a finite sum, contrary to the case of continues time, provides more accurate numerical results.

Definition 14.8. Let $L(k) := (L_{ij}(k); i, j \in E)$, be an unknown matrix-valued matrix and $G(k) := (G_{ij}(k); i, j \in E)$ be a known matrix-valued function. the equation

$$L(k) = G(k) + q * L(k), \quad k \in \mathbb{N},$$

is called a discrete-time Markov renewal equation.

It has been proved (Proposition 5 [3]) that the Markov renewal equation has a unique solution

$$L(k) = \psi * G(k), \quad k \in \mathbb{N}.$$

14.2.2 Semimartingales

We give now a short introduction to the martingale and semimartingale theory (see, e.g., Jacod and Shiryaev [7]). Let us consider a complete probability space $(\Omega, \mathcal{F}, \mathbb{P})$. On continuous time, we can define the filtraton $\mathbf{F} = (\mathcal{F}_t)_{t \in \mathbb{R}_+}$ with respect to a nonempty space Ω contained in $\mathcal{F}$, i.e., an increasing and right-continuous family of sub-σ-algebras of $\mathcal{F}$. Moreover, we assume that $\mathcal{F}_0$ contains all $\mathbb{P}$-null sets of $\mathcal{F}$. Consider the continuous-time filtered probability space with respect to filtration **F** as the quadruple $\mathcal{B} = (\Omega, \mathcal{F}, \mathbf{F}, \mathbb{P})$. For each $n \in \mathbb{N}$, let $\mathcal{B}^n = (\Omega^n, \mathcal{F}^n, \mathbf{F}^n, \mathbb{P}^n)$, $\mathbf{F}^n = (\mathcal{F}^n_t)_{t \in \mathbb{R}_+}$, be a stochastic basis. In our case, consider the continuous-time filtration $\mathcal{F}^n_t = \mathcal{F}_{N(\lfloor nt \rfloor)} = \sigma(J_0, J_i, X_i; 1 \le i \le N(\lfloor nt \rfloor))$ with $\mathcal{F}_0 = \sigma(J_0)$. On discrete time, the filtration is denoted as $\tilde{\mathbf{F}} = (\mathcal{F}_k)_{k \in \mathbb{N}}$. Notice that the right-continuity has no meaning here. Therefore, the discrete-time filtered probability space with respect to filtration $\tilde{\mathbf{F}}$ is accordingly the quadruple $\tilde{\mathcal{B}} = (\Omega, \mathcal{F}, \tilde{\mathbf{F}}, \mathbb{P})$. Respectively, consider, for all $n \in \mathbb{N}$, a discrete-time stochastic basis $\tilde{\mathcal{B}}^n = (\Omega^n, \mathcal{F}^n, \tilde{\mathbf{F}}^n, \mathbb{P}^n)$. We assume that, for any $n \in \mathbb{N}^*$, it holds $\mathcal{F}^n_k = \mathcal{F}_k = \sigma(J_0, J_i, X_i; 1 \le i \le k)$.

Definition 14.9. 1. For each $n \in \mathbb{N}^*$, a martingale is defined as an adapted process $(U^n_t)_{t \in \mathbb{R}_+}$, $n \in \mathbb{N}^*$, on the filtered probability space $\mathcal{B}^n$ whose $\mathbb{P}$-almost all paths are càdlàg such that every U^n_t is integrable and that $\mathbb{E}[U^n_t | \mathcal{F}^n_s] = U^n_s$, $s \le t$. A discrete-time martingale is defined as an adapted process $(U^n_m)_{m \in \mathbb{N}}$, $n \in \mathbb{N}^*$, on the space $\tilde{\mathcal{B}}$ such that $\mathbb{E}[U^n_m | \mathcal{F}^n_l] = U^n_l$, $l \le m$.
2. An $\mathbb{R}^d$-valued martingale $(U^n_t)_{t \in \mathbb{R}_+}$, $n \in \mathbb{N}^*$, is said to be a Gaussian martingale on $\mathcal{B}^n$ if $U^n_0 = 0$ and the distribution of any finite family $(U^n_{t_1}, \ldots, U^n_{t_k})$ is Gaussian. Every Gaussian martingale is characterized by a triple $(v, \mathcal{C}, B)$.
3. A stochastic process $(U^n_m)_{m \in \mathbb{N}^*}$, $n \in \mathbb{N}^*$, is a martingale difference on $\tilde{\mathcal{B}}^n$, if its expectation with respect to the filtration $\tilde{\mathbf{F}}$, is zero, i.e., $\mathbb{E}[U^n_m | \mathcal{F}^n_{m-1}] = 0$, for all $n \in \mathbb{N}^*$.
4. A semimartingale defined on the filtered probability space $\mathcal{B}^n$, as an adapted process $(X^n_t)_{t \in \mathbb{R}_+}$, $n \in \mathbb{N}^*$, that it can be decomposed as $X^n_t = M^n_t + A^n_t$, where $(M^n_t)_{t \in \mathbb{R}_+}$, $n \in \mathbb{N}^*$, is a local martingale and $(A^n_t)_{t \in \mathbb{R}_+}$, $n \in \mathbb{N}^*$, a càdlàg-adapted process of local bounded variation.

Conditional Lindeberg Condition : Let, for each $n \in \mathbb{N}^*$, $(U^n_m)_{m \in \mathbb{N}^*}$ be a martingale difference on $\tilde{\mathcal{B}}^n$. For any $\varepsilon > 0$, $t \in \mathbb{R}^*_+$, we have

$$\sum_{m=1}^{N(nt)} \mathbb{E}[\|U^n_m\|^2 \mathbf{1}_{\{\|U^n_m\| > \varepsilon\}} | \mathcal{F}^n_{m-1}] \xrightarrow{\mathbf{P}} 0, \quad n \to \infty, \tag{14.8}$$

where $\| \cdot \|$ is the euclidean norm.

14.3 Basic Results

We consider the finite set $E = \{1, \ldots, d\}$, $d \in \mathbb{N}^*$, and an observation of an SMC (Z_k), with state space E, up to a fixed censoring time $k \in \mathbb{N}^*$ as follows:

$$\mathcal{H}_k := \{Z_u, 0 \le u \le k\} = \begin{cases} \{J_0, X_1, \ldots, J_{N(k)}, U_k\}, & \text{if } N(k) > 0, \\ \{J_0, U_k = k\}, & \text{if } N(k) = 0, \end{cases}$$

where $U_k := k - S_{N(k)}$ is the backward recurrence time.

The empirical estimator $\hat{q}(x,k) := (\hat{q}_{ij}(x,k); i,j \in E)$, $x \in \{0,\ldots,k\}$, $k \in \mathbb{N}^*$, of the DTSMK (14.2) is defined by the following equation:

$$\hat{q}_{ij}(x,k) := \frac{1}{N_i(k)} \sum_{n=1}^{N(k)} \mathbf{1}_{\{J_{n-1}=i,J_n=j,X_n=x\}}, \qquad (14.9)$$

Remark 14.10. We notice that the backward recurrence times are neglected by the empirical estimators. As t tends to infinity, U_k adds no significant information to the asymptotic properties of the estimators.

We denote by $\lfloor \alpha \rfloor$ the integer part of a positive real number α and by δ the Kronecker's delta, i.e., $\delta_{ij} = 1$, if $i = j$, and 0, if $i \neq j$.

For any states $i,j \in E$, any fixed time $x \in \mathbb{N}$, we set $Y_m = (Y_m^{ij}; i,j \in E) \in \mathbb{R}^{d^2}$, where $(Y_m^{ij})_{m \in \mathbb{N}^*}$ are random sequences on $\tilde{\mathcal{B}}$ defined by

$$Y_m^{ij} := \mathbf{1}_{\{J_{m-1}=i,J_m=j,X_m=x\}} - \mathbf{1}_{\{J_{m-1}=i\}} q_{ij}(x), \qquad (14.10)$$

Let us define the double sequence $(Y_m^n)_{m \in \mathbb{N}^*}$, $n \in \mathbb{N}^*$, on $\tilde{\mathcal{B}}^n$, by

$$Y_m^n := \frac{Y_m}{\sqrt{n}}, \qquad (14.11)$$

Also, for any $t \in \mathbb{R}_+^*$, we denote the stochastic processes

$$S_t^{n,ij} := \sum_{m=1}^{N(\lfloor nt \rfloor)} Y_m^{n,ij} = \frac{1}{\sqrt{n}} \sum_{m=1}^{N(\lfloor nt \rfloor)} Y_m^{ij}, \quad n \in \mathbb{N}^*. \qquad (14.12)$$

Making use of the notation above, we have the sequence of stochastic processes on $\mathcal{B}^n$

$$S_t^n = (S_t^{n,ij}; i,j \in E), \quad t \in \mathbb{R}_+^*, \ n \in \mathbb{N}^*. \qquad (14.13)$$

Our results hold in the Skorohord space $D[0,\infty)$. We denote by $\xrightarrow{a.s.}$ the almost sure convergence, by $\Rightarrow$ the weak convergence in this space and by $\xrightarrow{D}$ the convergence in distribution of a sequence of random variables. $W_t := (W_t^{ij}; i,j \in E)$ is the d^2-dimensional standard Wiener process.

Lemma 14.11. *The conditional Lindeberg condition (14.8) holds for the random sequences $(Y_m^n)_{m \in \mathbb{N}^*}$, $n \in \mathbb{N}^*$, defined in (14.10),(14.11), i.e., for any $\varepsilon > 0$ and any $t \in \mathbb{R}_+^*$,*

$$\frac{1}{n} \sum_{m=1}^{N(\lfloor nt \rfloor)} \mathbb{E}[\|Y_m\|^2 \mathbf{1}_{\{\|Y_m\|>\varepsilon\sqrt{n}\}} | \mathcal{F}_{m-1}] \xrightarrow{a.s.} 0, \quad n \to \infty,$$

where $\|\cdot\|$ is the euclidean norm on $\mathbb{R}^{d^2}$.

Proof. First, we must show that the random sequences $(Y_m^n)_{m\in\mathbb{N}^*}$, $n\in\mathbb{N}^*$, are $(\mathcal{F}_m)$-martingale differences on $\tilde{\mathcal{B}}^n$. For any $i,j\in E$ and any fixed $x\in\mathbb{N}^*$, we have

$$
\begin{aligned}
\mathbb{E}[Y_m^{ij}|\mathcal{F}_{m-1}] &= \mathbb{E}[\mathbf{1}_{\{J_{m-1}=i,J_m=j,X_m=x\}} - \mathbf{1}_{\{J_{m-1}=i\}}q_{ij}(x)|\mathcal{F}_{m-1}]\\
&= \mathbb{E}[\mathbf{1}_{\{J_{m-1}=i,J_m=j,X_m=x\}}|\mathcal{F}_{m-1}] - \mathbb{E}[\mathbf{1}_{\{J_{m-1}=i\}}q_{ij}(x)|\mathcal{F}_{m-1}]\\
&= \mathbb{E}[\mathbf{1}_{\{J_{m-1}=i,J_m=j,X_m=x\}}|J_{m-1}] - \mathbb{E}[\mathbf{1}_{\{J_{m-1}=i\}}q_{ij}(x)|J_{m-1}]\\
&= \mathbf{1}_{\{J_{m-1}=i\}}\mathbb{P}(J_m=j,X_m=x|J_{m-1}) - \mathbf{1}_{\{J_{m-1}=i\}}q_{ij}(x)\\
&= \mathbf{1}_{\{J_{m-1}=i\}}q_{ij}(x) - \mathbf{1}_{\{J_{m-1}=i\}}q_{ij}(x) = 0.
\end{aligned}
$$

So, for any $n\in\mathbb{N}^*$, $(Y_m^n)_{m\in\mathbb{N}^*}$ are $(\mathcal{F}_m)$-martingale differences. As a direct outcome, the processes $(S_t^n)_{t\in\mathbb{R}_+^*}$ is a martingale on $\mathcal{B}^n$.

Now, for all $i,j\in E$ and all $m\in\mathbb{N}$, $|Y_m^{ij}|\le 1$. So, $\|Y_m\|\le d$. Using Markov's inequality, for any $\varepsilon>0$ and any $t\in\mathbb{R}_+^*$, we get

$$
\begin{aligned}
\frac{1}{n}\sum_{m=1}^{N(\lfloor nt\rfloor)}\mathbb{E}[\|Y_m\|^2\mathbf{1}_{\{\|Y_m\|>\varepsilon\sqrt{n}\}}|\mathcal{F}_{m-1}] &\le \frac{1}{n}\sum_{m=1}^{N(\lfloor nt\rfloor)}\mathbb{E}[d^2\mathbf{1}_{\{\|Y_m\|>\varepsilon\sqrt{n}\}}|\mathcal{F}_{m-1}]\\
&= \frac{d^2}{n}\sum_{m=1}^{N(\lfloor nt\rfloor)}\mathbb{E}[\mathbf{1}_{\{\|Y_m\|>\varepsilon\sqrt{n}\}}|\mathcal{F}_{m-1}]\\
&= \frac{d^2}{n}\sum_{m=1}^{N(\lfloor nt\rfloor)}\mathbb{P}(\|Y_m\|>\varepsilon\sqrt{n}|\mathcal{F}_{m-1})\\
&\le \frac{d^2}{n}\sum_{m=1}^{N(\lfloor nt\rfloor)}\frac{\mathbb{E}[\|Y_m\||\mathcal{F}_{m-1}]}{\varepsilon\sqrt{n}}\\
&\le \frac{d^2}{n}\sum_{m=1}^{N(\lfloor nt\rfloor)}\frac{d}{\varepsilon\sqrt{n}}\\
&= \frac{d^3}{\varepsilon\,n\sqrt{n}}N(\lfloor nt\rfloor)\xrightarrow{a.s.}0,\quad n\to\infty,
\end{aligned}
$$

since, by Lemma 14.4, $N(\lfloor nt\rfloor)/n\xrightarrow{a.s.}t/\bar{m}$, as $n\to\infty$.   □

Remark 14.12. In Lemma 14.11, we have the stronger *a.s.* convergence instead of the convergence in probability required in the conditional Lindeberg condition.

Lemma 14.13. *For any $t\in\mathbb{R}_+^*$, the following a.s. convergence holds:*

$$
\frac{1}{n}\sum_{m=1}^{N(\lfloor nt\rfloor)}\mathbb{E}[Y_m^{ij}Y_m^{lr}|\mathcal{F}_{m-1}]\xrightarrow{a.s.}t\,C^{ij,lr},\quad n\to\infty,\tag{14.14}
$$

where

$$c^{ij,lr} = \delta_{il}\frac{1}{\mu_{ii}}q_{ij}(x)(\delta_{ir} - q_{ir}(x)), \quad (i,j),(l,r) \in E \times E.$$

Proof. We distinguish the following cases:

1. For $i \neq l$, we have $Y_m^{ij}Y_m^{lr} = 0$. Consequently,

$$\mathbb{E}[Y_m^{ij}Y_m^{lr}|\mathcal{F}_{m-1}] = 0. \tag{14.15}$$

2. For $i = l$ and $j \neq r$, for any $x \in \mathbb{N}$

$$Y_m^{ij}Y_m^{lr} = Y_m^{ij}Y_m^{ir} = -\mathbf{1}_{\{J_{m-1}=i,J_m=j,X_m=x\}}q_{ir}(x)$$
$$-\mathbf{1}_{\{J_{m-1}=i,J_m=r,X_m=x\}}q_{ij}(x)$$
$$+\mathbf{1}_{\{J_{m-1}=i\}}q_{ij}(x)q_{ir}(x).$$

So, we get that

$$\mathbb{E}[Y_m^{ij}Y_m^{lr}|\mathcal{F}_{m-1}] = -q_{ir}(x)\mathbb{P}(J_{m-1} = i, J_m = j, X_m = x|J_{m-1})$$
$$- q_{ij}(x)\mathbb{P}(J_{m-1} = i, J_m = r, X_m = x|J_{m-1})$$
$$+ q_{ij}(x)q_{ir}(x)\mathbb{P}(J_{m-1} = i|J_{m-1})$$
$$= -q_{ij}(x)q_{ir}(x)\mathbf{1}_{\{J_{m-1}=i\}}$$

Now, we can write

$$\frac{1}{n}\sum_{m=1}^{N(\lfloor nt \rfloor)}\mathbb{E}[Y_m^{ij}Y_m^{lr}|\mathcal{F}_{m-1}] = -\frac{1}{n}\sum_{m=1}^{N(\lfloor nt \rfloor)}\mathbf{1}_{\{J_{m-1}=i\}}q_{ij}(x)q_{ir}(x).$$

We mention that $\sum_{m=1}^{N(\lfloor nt \rfloor)}\mathbf{1}_{\{J_{m-1}=i\}} = N_i(\lfloor nt \rfloor)$. From Lemma 14.4, we conclude that

$$\frac{1}{n}\sum_{m=1}^{N(\lfloor nt \rfloor)}\mathbb{E}[Y_m^{ij}Y_m^{lr}|\mathcal{F}_{m-1}] \xrightarrow{a.s.} -\frac{t}{\mu_{ii}}q_{ij}(x)q_{ir}(x), \quad n \to \infty. \tag{14.16}$$

3. For $i = l$ and $j = r$, for any $x \in \mathbb{N}$

$$Y_m^{ij}Y_m^{lr} = (Y_m^{ij})^2 = +\mathbf{1}_{\{J_{m-1}=i,J_m=j,X_m=x\}}$$
$$+\mathbf{1}_{\{J_{m-1}=i\}}q_{ij}^2(x)$$
$$-2\mathbf{1}_{\{J_{m-1}=i,J_m=j,X_m=x\}}q_{ij}(x).$$

Consequently,

$$\mathbb{E}[Y_m^{ij}Y_m^{lr}|\mathcal{F}_{m-1}] = \mathbb{P}(J_{m-1} = i, J_m = j, X_m = x|J_{m-1})$$
$$+ q_{ij}^2(x)\mathbb{P}(J_{m-1} = i|J_{m-1})$$
$$- 2q_{ij}(x)\mathbb{P}(J_{m-1} = i, J_m = j, X_m = x|J_{m-1})$$
$$= \mathbf{1}_{\{J_{m-1}=i\}}q_{ij}(x)(1 - q_{ij}(x))$$

Now,

$$\frac{1}{n}\sum_{m=1}^{N(\lfloor nt \rfloor)} \mathbb{E}[Y_m^{ij}Y_m^{lr}|\mathcal{F}_{m-1}] = \frac{1}{n}\sum_{m=1}^{N(\lfloor nt \rfloor)} \mathbf{1}_{\{J_{m-1}=i\}}q_{ij}(x)(1 - q_{ij}(x)).$$

From Lemma 14.4, we have

$$\frac{1}{n}\sum_{m=1}^{N(\lfloor nt \rfloor)} \mathbb{E}[Y_m^{ij}Y_m^{lr}|\mathcal{F}_{m-1}] \xrightarrow{a.s.} \frac{t}{\mu_{ii}}q_{ij}(x)(1 - q_{ij}(x)), \quad n \to \infty. \qquad (14.17)$$

So, from (14.15), (14.16), and (14.17), the *a.s.* convergence (14.14) holds, where

$$C^{ij,lr} = \begin{cases} \frac{1}{\mu_{ii}}q_{ij}(x)(1 - q_{ij}(x)), & \text{if } i = l \text{ and } j = r, \\ -\frac{1}{\mu_{ii}}q_{ij}(x)q_{ir}(x), & \text{if } i = l \text{ and } j \neq r, \\ 0, & \text{if } i \neq l. \end{cases}$$

$\square$

Consider the $d^2 \times d^2$-dimensional matrix $\mathcal{C} = (C^{ij,lr})_{(i,j),(l,r)\in E\times E}$. It may be more convenient to present the matrix $\mathcal{C}$ as a diagonal matrix in block form

$$\mathcal{C} = \begin{pmatrix} \mathcal{C}_1 & \mathbf{0} & \cdots & \mathbf{0} \\ \mathbf{0} & \mathcal{C}_2 & \cdots & \mathbf{0} \\ \vdots & \vdots & \ddots & \vdots \\ \mathbf{0} & \mathbf{0} & \cdots & \mathcal{C}_d \end{pmatrix}$$

with $\mathcal{C}_i$, for any $i \in E$, to be the block

$$\mathcal{C}_i = \frac{1}{\mu_{ii}}q_{ij}(x)(\delta_{jr} - q_{ir}(x)), \quad j,r \in E$$

$$= \frac{1}{\mu_{ii}}\begin{pmatrix} q_{i1}(x)(1-q_{i1}(x)) & -q_{i1}(x)q_{i2}(x) & -q_{i1}(x)q_{i3}(x) & \cdots & -q_{i1}(x)q_{id}(x) \\ -q_{i2}(x)q_{i1}(x) & q_{i2}(x)(1-q_{i2}(x)) & -q_{i2}(x)q_{i3}(x) & \cdots & -q_{i2}(x)q_{id}(x) \\ -q_{i3}(x)q_{i1}(x) & -q_{i3}(x)q_{i2}(x) & q_{i3}(x)(1-q_{i3}(x)) & \cdots & -q_{i3}(x)q_{id}(x) \\ \vdots & \vdots & \vdots & \ddots & \vdots \\ -q_{id}(x)q_{i1}(x) & -q_{id}(x)q_{i2}(x) & -q_{id}(x)q_{i3}(x) & \cdots & q_{id}(x)(1-q_{id}(x)) \end{pmatrix}.$$

Lemma 14.14. *The $d^2 \times d^2$-dimensional matrix C is a covariance matrix, i.e., it is symmetric and positive semi-definite.*

Proof. Every block of C is symmetric and, consequently, the whole matrix C is also symmetric. To show that it is positive semi-definite, it suffices to prove that

$$z^\top C z = \sum_{i,j=1}^{s} \sum_{l,r=1}^{s} z_{ij} C^{ij,lr} z_{lr} \geq 0,$$

for any nonzero $z = [z_{11} z_{12} \cdots z_{1d} z_{21} z_{22} \cdots z_{2d} \cdots z_{d1} z_{d2} \cdots z_{dd}]^\top \in \mathbb{R}^{d^2}$. Thus,

$$z^\top C z = \sum_{i \in E} \left[\frac{t}{\mu_{ii}} \sum_{j,r \in E} q_{ij}(x)(\delta_{jr} - q_{ir}(x)) z_{ij} z_{ir} \right]$$

$$= \sum_{i \in E} \frac{t}{\mu_{ii}} \sum_{j \in E} \left[q_{ij}(x)(1 - q_{ij}(x)) z_{ij}^2 - \sum_{\substack{r \in E \\ j \neq r}} q_{ij}(x) q_{ir}(x) z_{ij} z_{ir} \right].$$

For any $x \in \mathbb{N}$, we have that $\sum_{j \in E} q_{ij}(x) \leq 1$. That is, for any $i \in E$,

$$1 - q_{il}(x) \geq \sum_{\substack{j \in E \\ j \neq l}} q_{ij}(x).$$

Moreover, we take advantage of the symmetry of every block of C and aggregate the elements of the second part in the parenthesis. So, we get that

$$z^\top C z \geq \sum_{i \in E} \frac{1}{\mu_{ii}} \sum_{j \in E} \left[q_{ij}(x) \sum_{\substack{l \in E \\ l \neq j}} q_{il}(x) z_{ij}^2 - 2 \sum_{\substack{r \in E \\ r > j}} q_{ij}(x) q_{ir}(x) z_{ij} z_{ir} \right].$$

After some computations, we derive that

$$z^\top C z \geq \sum_{i \in E} \frac{t}{\mu_{ii}} \sum_{\substack{j,r \in E \\ r > j}} q_{ij}(x) q_{ir}(x)(z_{ij} - z_{ir})^2 \geq 0.$$

So, the matrix C is a covariance matrix.   □

Theorem 14.15. *Let S_t^n be the processes defined in (14.12),(14.13). Then, for any fixed $x \in \mathbb{N}$, the following weak convergence holds:*

$$(S_t^n; t \in \mathbb{R}_+^*) \Rightarrow (C^{1/2} W_t; t \in \mathbb{R}_+^*), \quad n \to \infty,$$

where

$$C = (C^{il,lr})_{(i,j),(l,r) \in E \times E} = \left(\delta_{il} \frac{1}{\mu_{ii}} q_{ij}(x)(\delta_{jr} - q_{ir}(x)) \right)_{(i,j),(l,r) \in E \times E}$$

is the covariance matrix.

Proof. We recall that $(S_t^n)_{t \in \mathbb{R}_+^*}$ is a martingale on $\mathcal{B}^n$ and therefore a semimartingale. Following Lemmas 14.4, 14.11, 14.13 and 14.14, and from VIII Theorem 3.33, Jacod and Shiryaev [7], we get that

$$\left(S_t^n; t \in \mathbb{R}_+^* \right) \Rightarrow \left(S_t; t \in \mathbb{R}_+^* \right), \quad n \to \infty,$$

where $(S_t)_{t \in \mathbb{R}_+^*}$ is a d^2-dimensional continuous Gaussian martingale with predictable characteristics $(0, t\,C, 0)$.

A Gaussian martingale on the filtered probability space $\mathcal{B}^n$ with characteristics $(0, t\,C, 0)$ is a standard Brownian motion (II Theorem 4.36, Jacod and Shiryaev [7]). Furthermore, its variance function is $\sigma^2(x,t) = \langle W, W \rangle_t = t\,C$. So, we get the final result.   □

For Ξ a quantity to be estimated, we will denote by $\Delta\Xi$ the difference between the estimator of Ξ and the true value of Ξ. For instance, for any $i, j \in E$ and $x \in [0, \lfloor nt \rfloor]$, $n \in \mathbb{N}^*, t \in \mathbb{R}_+^*$, we set $\Delta q_{ij}(x, \lfloor nt \rfloor) := \hat{q}_{ij}(x, \lfloor nt \rfloor) - q_{ij}(x)$. Under this notation, we proceed to the following weak convergence theorem.

Theorem 14.16. *For any fixed $x \in \mathbb{N}$, it holds the weak convergence*

$$\left(\sqrt{n}\Delta q_{ij}(x, \lfloor nt \rfloor); i, j \in E, t \in \mathbb{R}_+^* \right) \Rightarrow \left(\mathcal{G}_{ij}(t); i, j \in E, t \in \mathbb{R}_+^* \right), \quad n \to \infty,$$

where $\mathcal{G}_{t,ij} = \mu_{ii}\sqrt{C^{ij,lr}} W_t^{ij}$ is a continuous Gaussian process.

Proof. The processes $\Delta q_{ij}(x, nt)$, $x \in [0, nt]$, $n \in \mathbb{N}^*, t \in \mathbb{R}_+^*$, can be written as

$$\Delta q_{ij}(x, \lfloor nt \rfloor) = \frac{1}{N_i(\lfloor nt \rfloor)} \sum_{m=1}^{N(\lfloor nt \rfloor)} \left[\mathbf{1}_{\{J_{m-1}=i, J_m=j, X_m=x\}} - \mathbf{1}_{\{J_{m-1}=i\}} q_{ij}(x) \right]$$

$$= \frac{1}{N_i(\lfloor nt \rfloor)} \sum_{m=1}^{N(\lfloor nt \rfloor)} Y_m^{ij}, \quad i, j \in E.$$

So, we can write

$$\left(\sqrt{n}\Delta q_{ij}(x, nt); i, j \in E \right) = \left(\frac{\sqrt{n}}{N_i(nt)} \sum_{k=1}^{N(nt)} Y_k^{ij}; i, j \in E \right)$$

$$= \left(\frac{n}{N_i(nt)} \frac{1}{\sqrt{n}} \sum_{k=1}^{N(nt)} Y_k^{ij}; i, j \in E \right)$$

$$= \left(\frac{n}{N_i(nt)} S_t^{n,ij}; i, j \in E \right).$$

From Theorem 14.15 and the Slutsky Lemma, we conclude that

$$(\sqrt{n}\Delta q_{ij}(x,\lfloor nt \rfloor); i,j \in E, t \in \mathbb{R}_+^*) \Rightarrow \left(\mu_{ii}\sqrt{C^{ij,lr}}\frac{W_t^{ij}}{t}; i,j \in E, t \in \mathbb{R}_+^*\right), \quad n \to \infty,$$

since by Lemma 14.4, $N_i(\lfloor nt \rfloor)/n \xrightarrow{a.s.} t/\mu_{ii}$, as $n \to \infty$. It is easy to see that, for $t \in \mathbb{R}_+^*$, $\mathcal{G}_{t,ij} = \mu_{ii}\sqrt{C^{ij,lr}}W_t^{ij}$ is a continuous Gaussian process. □

Remark 14.17. In fact, the result of Theorem 14.16 is the invariance principle in multidimensional form for the empirical estimator (14.9) of a DTSMK (14.2).

Corollary 14.18. *For any fixed $x \in \mathbb{N}$ we have*

$$(\sqrt{n}\Delta q_{ij}(x,k); i,j \in E) \xrightarrow{D} \mathcal{N}(\mathbf{0},\Sigma_q), \quad k \to \infty,$$

where $\mathcal{N}(\mathbf{0},\Sigma_q)$ is a d^2-dimensional normal random variable and $\Sigma_q = (\mu_{ii}^2 C^{ij,lr})_{(i,j),(l,r)\in E\times E}$

Proof. Setting $t = 1$, we have that $W_1 \sim \mathcal{N}(0,1)$. Then, we replace n with k. As $\mathcal{G}_{1,ij} = \mu_{ii}\sqrt{C^{ij,lr}}W_1^{ij}$, we have the desired result. □

14.4 Asymptotics for Kernels and Functionals

We can verify all the necessary conditions as described in Lemmas 14.11, 14.13, and 14.14 for the following kernels and functionals. We give the empirical estimators and then we present directly the corresponding weak convergence for these measures without proofs. For simplicity, some symbols have not been not changed.

14.4.1 Transition Kernel

The empirical estimator $\hat{P}(k) = (\hat{p}_{ij}(k); i,j \in E)$, $k \in \mathbb{N}$, of the transition kernel (14.1) is defined by

$$\hat{p}_{ij}(k) := \frac{1}{N_i(k)} \sum_{m=1}^{N(k)} \mathbf{1}_{\{J_{m-1}=i,J_m=j\}},$$

For any states $i,j \in E$, we set $Y_m = (Y_m^{ij}; i,j \in E) \in \mathbb{R}^{d^2}$, where $(Y_m^{ij})m \in \mathbb{N}^*$ are random sequences on $\tilde{\mathcal{B}}$ defined by

$$Y_m^{ij} := \mathbf{1}_{\{J_{m-1}=i,J_m=j\}} - \mathbf{1}_{\{J_{m-1}=i\}}p_{ij},$$

Theorem 14.19. *The following weak convergence holds.*

$$\left(\sqrt{n}\Delta p_{ij}(\lfloor nt \rfloor); i, j \in E, t \in \mathbb{R}_+^*\right) \Rightarrow \left(\mathcal{G}_{ij}(t); i, j \in E, t \in \mathbb{R}_+^*\right), \quad n \to \infty,$$

where $\mathcal{G}_{t,ij} = \mu_{ii}\sqrt{\mathcal{C}^{ij,lr}}W_t^{ij}$ *is a continuous Gaussian process with*

$$\mathcal{C} = (\mathcal{C}^{ij,lr})_{(i,j),(l,r)\in E\times E} = \left(\delta_{il}\frac{1}{\mu_{ii}}p_{ij}(\delta_{ir} - p_{ir})\right)_{(i,j),(l,r)\in E\times E}.$$

Corollary 14.20. *The following convergence holds:*

$$\left(\sqrt{n}\Delta p_{ij}(k); i, j \in E\right) \xrightarrow{D} \mathcal{N}(\mathbf{0}, \Sigma_P), \quad k \to \infty,$$

where $\mathcal{N}(\mathbf{0}, \Sigma_P)$ *is a* d^2*-dimensional normal random variable and* $\Sigma_P = (\mu_{ii}^2 \mathcal{C}^{ij,lr})_{(i,j),(l,r)\in E\times E}$.

14.4.2 Cumulative DTSMK

The empirical estimator $\hat{Q}(x,k) = (\hat{Q}_{ij}(x,k); i, j \in E)$, $x \in \{0,\ldots,k\}$, $k \in \mathbb{N}$, of the cumulative DTSMK (14.3) is defined by

$$\hat{Q}_{ij}(x,k) := \frac{1}{N_i(k)}\sum_{m=1}^{N(k)}\mathbf{1}_{\{J_{m-1}=i,J_m=j,X_m\leq x\}},$$

For any states $i, j \in E$, any fixed time $x \in \mathbb{N}$, we set $Y_m = (Y_m^{ij}; i, j \in E) \in \mathbb{R}^{d^2}$, where $(Y_m^{ij})_{m\in\mathbb{N}^*}$ are random sequences on $\tilde{\mathcal{B}}$ defined by

$$Y_m^{ij} := \mathbf{1}_{\{J_{m-1}=i,J_m=j,X_m\leq x\}} - \mathbf{1}_{\{J_{m-1}=i,J_m=j\}}Q_{ij}(x),$$

Theorem 14.21. *For any fixed* $x \in \mathbb{N}$ *it holds the weak convergence*

$$\left(\sqrt{n}\Delta Q_{ij}(x, \lfloor nt \rfloor); i, j \in E, t \in \mathbb{R}_+^*\right) \Rightarrow \left(\mathcal{G}_{ij}(t); i, j \in E, t \in \mathbb{R}_+^*\right), \quad n \to \infty,$$

where $\mathcal{G}_{t,ij} = \mu_{ii}\mathcal{C}^{1/2}W_t/t$ *is a continuous Gaussian process with*

$$\mathcal{C} = (\mathcal{C}^{ij,lr})_{(i,j),(l,r)\in E\times E} = \left(\delta_{il}\frac{1}{\mu_{ii}}Q_{ij}(x)(\delta_{ir} - Q_{ir}(x))\right)_{(i,j),(l,r)\in E\times E}.$$

Corollary 14.22. *For any fixed* $x \in \mathbb{N}$, *we have*

$$\left(\sqrt{n}\Delta Q_{ij}(x,k); i, j \in E\right) \xrightarrow{D} \mathcal{N}(\mathbf{0}, \Sigma_Q), \quad k \to \infty,$$

where $\mathcal{N}(\mathbf{0}, \Sigma_Q)$ *is a* d^2*-dimensional normal random variable and* $\Sigma_Q = (\mu_{ii}^2 \mathcal{C}^{ij,lr})_{(i,j),(l,r)\in E\times E}$.

14.4.3   Conditional Sojourn Time Distribution Function

The empirical estimator $\hat{f}(x,k) = (\hat{f}_{ij}(x,k); i,j \in E)$, $x \in \{0,\ldots,k\}$, $k \in \mathbb{N}$, of the conditional sojourn time distribution function (14.4) is defined by

$$\hat{f}_{ij}(x,k) := \frac{1}{N_{ij}(k)} \sum_{m=1}^{N(k)} \mathbf{1}_{\{J_{m-1}=i, J_m=j, X_m=x\}},$$

For any states $i,j \in E$, any fixed time $x \in \mathbb{N}$, we set $Y_m = (Y_m^{ij}; i,j \in E) \in \mathbb{R}^{d^2}$, where $(Y_m^{ij})_{m \in \mathbb{N}^*}$ are random sequences on $\tilde{\mathcal{B}}$ defined by

$$Y_m^{ij} := \mathbf{1}_{\{J_{m-1}=i, J_m=j, X_m=x\}} - \mathbf{1}_{\{J_{m-1}=i, J_m=j\}} f_{ij}(x),$$

Theorem 14.23. *For any fixed $x \in \mathbb{N}$, it holds the convergence*

$$\left(\sqrt{n}\Delta f_{ij}(x, \lfloor nt \rfloor); i,j \in E, t \in \mathbb{R}_+^*\right) \Rightarrow \left(\mathcal{G}_{ij}(t); i,j \in E, t \in \mathbb{R}_+^*\right), \quad n \to \infty,$$

where $\mathcal{G}_{t,ij} = \frac{\mu_{ii}}{p_{ij}} \mathcal{C}^{1/2} W_t / t$ is a continuous Gaussian process with

$$\mathcal{C} = (\mathcal{C}^{ij,lr})_{(i,j),(l,r) \in E \times E} = \left(\delta_{il}\delta_{ir}\frac{1}{\mu_{ii}}q_{ij}(x)(1 - f_{ij}(x))\right)_{(i,j),(l,r) \in E \times E}.$$

Corollary 14.24. *For any fixed $x \in \mathbb{N}$, we have*

$$\left(\sqrt{n}\Delta f_{ij}(x,k); i,j \in E\right) \xrightarrow{D} \mathcal{N}(0, \Sigma_f), \quad k \to \infty,$$

where $\mathcal{N}(0, \Sigma_f)$ is a d^2-dimensional normal random variable and $\Sigma_f = \left(\frac{\mu_{ii}^2}{p_{ij}^2}\mathcal{C}^{ij,lr}\right)_{(i,j),(l,r) \in E \times E}$.

14.4.4   Conditional Sojourn Time Cumulative Distribution Function

The empirical estimator $\hat{F}(x,k) = (\hat{F}_{ij}(x,k); i,j \in E)$, $x \in \{0,\ldots,k\}$, $k \in \mathbb{N}$, of the conditional sojourn time cumulative distribution function (14.5) is defined by

$$\hat{F}_{ij}(x,k) := \frac{1}{N_{ij}(k)} \sum_{m=1}^{N(k)} \mathbf{1}_{\{J_{m-1}=i, J_m=j, X_m \leq x\}},$$

For any states $i,j \in E$, any fixed time $x \in \mathbb{N}$, we set $Y_m = (Y_m^{ij}; i,j \in E) \in \mathbb{R}^{d^2}$, where $(Y_m^{ij})_{m \in \mathbb{N}^*}$ are random sequences on $\tilde{\mathcal{B}}$ defined by

$$Y_m^{ij} := \mathbf{1}_{\{J_{m-1}=i, J_m=j, X_m \leq x\}} - \mathbf{1}_{\{J_{m-1}=i, J_m=j\}} F_{ij}(x),$$

Theorem 14.25. *For any fixed $x \in \mathbb{N}$, it holds the convergence*

$$\left(\sqrt{n}\Delta F_{ij}(x, \lfloor nt \rfloor); i, j \in E, t \in \mathbb{R}_+^*\right) \Rightarrow \left(\mathcal{G}_{ij}(t); i, j \in E, t \in \mathbb{R}_+^*\right), \quad n \to \infty,$$

where $\mathcal{G}_{t,ij} = \frac{\mu_{ii}}{p_{ij}}\mathcal{C}^{1/2}W_t/t$ is a continuous Gaussian process with

$$\mathcal{C} = (\mathcal{C}^{ij,lr})_{(i,j),(l,r)\in E \times E} = \left(\delta_{il}\delta_{ir}\frac{1}{\mu_{ii}}Q_{ij}(x)\left(1 - F_{ij}(x)\right)\right)_{(i,j),(l,r)\in E \times E}.$$

Corollary 14.26. *For any fixed $x \in \mathbb{N}$, we have*

$$(\sqrt{n}\Delta F_{ij}(x,k); i, j \in E) \xrightarrow{\mathcal{D}} \mathcal{N}(\mathbf{0}, \Sigma_F), \quad k \to \infty,$$

where $\mathcal{N}(\mathbf{0}, \Sigma_F)$ is a d^2-dimensional normal random variable and $\Sigma_F = \left(\frac{\mu_{ii}^2}{p_{ij}^2}\mathcal{C}^{ij,lr}\right)_{(i,j),(l,r)\in E \times E}$.

14.4.5   Sojourn Time Distribution Function

The empirical estimator $\hat{h}(x,k) = (\hat{h}_i(x,k); i \in E)$, $x \in \{0,\ldots,k\}$, $k \in \mathbb{N}$, of the sojourn time distribution function (14.6) is defined by

$$\hat{h}_i(x,k) := \frac{1}{N_i(k)}\sum_{m=1}^{N(k)}\mathbf{1}_{\{J_{m-1}=i, X_m=x\}},$$

For any states $i \in E$, any fixed time $x \in \mathbb{N}$, we set $Y_m = (Y_m^i; i \in E) \in \mathbb{R}^d$, where $(Y_m^i)_{m \in \mathbb{N}^*}$ are random sequences on $\tilde{\mathcal{B}}$ defined by

$$Y_m^i := \mathbf{1}_{\{J_{m-1}=i, X_m=x\}} - \mathbf{1}_{\{J_{m-1}=i\}}h_i(x),$$

Theorem 14.27. *For any fixed $x \in \mathbb{N}$, it holds the convergence*

$$\left(\sqrt{n}\Delta h_i(x, \lfloor nt \rfloor); i \in E, t \in \mathbb{R}_+^*\right) \Rightarrow \left(\mathcal{G}_i(t); i \in E, t \in \mathbb{R}_+^*\right), \quad n \to \infty,$$

where $\mathcal{G}_{t,i} = \mu_{ii}\mathcal{C}^{1/2}W_t/t$ is a continuous Gaussian process with

$$\mathcal{C} = (\mathcal{C}^{i,l})_{i,l\in E} = \left(\delta_{il}\frac{1}{\mu_{ii}}h_i(x)(1 - h_i(x))\right)_{i,l\in E}.$$

Corollary 14.28. *For any fixed $x \in \mathbb{N}$, we have*

$$(\sqrt{n}\Delta h_i(x,k); i \in E) \xrightarrow{\mathcal{D}} \mathcal{N}(\mathbf{0}, \Sigma_h), \quad t \to \infty,$$

where $\mathcal{N}(\mathbf{0}, \Sigma_h)$ is a d^2-dimensional normal random variable and $\Sigma_h = (\mu_{ii}^2\mathcal{C}^{i,l})_{i,l\in E}$.

14.4.6 Sojourn Time Cumulative Distribution Function

The empirical estimator $\hat{H}(x,k) = (\hat{h}_i(x,k); i \in E), x \in \{0,\ldots,k\}, k \in \mathbb{N}, k \in \mathbb{N}$, of the sojourn time cumulative distribution function (14.7) is defined by

$$\hat{H}_i(x,k) := \frac{1}{N_i(k)} \sum_{m=1}^{N(k)} \mathbf{1}_{\{J_{m-1}=i, X_m \leq x\}},$$

For any states $i \in E$, any fixed time $x \in \mathbb{N}$, we set $Y_m = (Y_m^i; i \in E) \in \mathbb{R}^d$, where $(Y_m^i)_{m \in \mathbb{N}^*}$ are random sequences on $\tilde{\mathcal{B}}$ defined by

$$Y_m^i := \mathbf{1}_{\{J_{m-1}=i, X_m \leq x\}} - \mathbf{1}_{\{J_{m-1}=i\}} H_i(x),$$

Theorem 14.29. *For any fixed $x \in \mathbb{N}$, it holds the convergence*

$$\left(\sqrt{n}\Delta H_i(x, \lfloor nt \rfloor); i \in E, t \in \mathbb{R}_+^*\right) \Rightarrow \left(\mathcal{G}_i(t); i \in E, t \in \mathbb{R}_+^*\right), \quad n \to \infty,$$

where $\mathcal{G}_{t,i} = \mu_{ii} C^{1/2} W_t / t$ is a continuous Gaussian process with

$$C = (C^{i,l})_{i,l \in E} = \left(\delta_{il} \frac{1}{\mu_{ii}} H_i(x)(1 - H_i(x))\right)_{i,l \in E}.$$

Corollary 14.30. *For any fixed $x \in \mathbb{N}$, we have*

$$\left(\sqrt{n}\Delta H_i(x,k); i \in E\right) \xrightarrow{\mathcal{D}} \mathcal{N}(0, \Sigma_H), \quad t \to \infty,$$

where $\mathcal{N}(0, \Sigma_H)$ is a d^2-dimensional normal random variable and $\Sigma_H = (\mu_{ii}^2 C^{i,l})_{i,l \in E}$.

Conclusion

The DTSMK constitute a key role in the probabilistic and statistical study of a SMC as the majority of the relative measures can be expressed directly in terms of it. Consequently, the thorough study of the DTSMK is of foremost importance in the further analysis of SMC.

The multidimensional invariance principle presented in this chapter is an interesting theoretical result oriented to applications. It seems rather useful for estimating various kernels and functions, observing a semi-Markov system up to fixed censoring time. For exemple, in dependability theory, we are interested in estimating measures such as the reliability, availability or failure rate functions, which are directly written in function of the DTSMK.

References

1. Barbu, V.S., Limnios, N.: Empirical estimation for discrete-time semi-Markov processes with applications in reliability. Nonparametric Stat. **8**(78), 483–498 (2006)
2. Barbu, V.S., Limnios, N.: Semi-Markov Chains and Hidden Semi-Markov Models Toward Applications. Springer, New York (2008)
3. Barbu, V.S., Boussemart, M., Limnios, N.: Discrete-time semi-Markov model for reliability and survival analysis. Comm. Stat. Theor. Meth. **33**(11), 2833–2868 (2004)
4. Georgiadis, S., Limnios, N.: A multidimensional functional central limit theorem for an empirical estimator of a continuous-time semi-Markov kernel. Nonparametric Stat. (submitted)
5. Gill, R.D.: Nonparametric estimation based on censored observations of a Markov renewal process, Z. Wahrscheinlichkeitstheorie verw. Gebiete, **53**, 97–116 (1980)
6. Greenwood, P.E., Wefelmeyer, W.: Empirical estimators for semi-Markov processes. Math. Methods Statist. **5**(3), 299–315 (1996)
7. Jacod, J., Shiryaev, A.: Limit Theorems for Stochastic Processes. 2nd edn, Springer, Berlin (2003)
8. Limnios, N.: A functional central limit theorem for the empirical estimator of a semi-Markov kernel. Nonparametric Stat. **16**(1-2), 13–18 (2004)
9. Limnios, N., Oprişan, G.: Semi-Markov Processes and Reliability. Birkhäuser, Boston (2004)
10. Moore, E.H., Pyke, R.: Estimation of the transition distributions of a Markov renewal process. Ann. Inst. Stat. Math. **20**, 411–424 (1968)
11. Ouhbi, B., Limnios, N.: Nonparametric estimation for semi-Markov processes based on its hazard rate functions. Stat. Infer. Stoch. Process. **2**, 151–173 (1999)
12. Ouhbi, B., Limnios, N.: Nonparametric reliability estimation of semi-Markov processes. J. Stat. Plann. Infer. **109**, 155–165 (2003)
13. Pyke, R.: Markov renewal processes: definitions and preliminary properties. Ann. Math. Stat. **32**, 1231–1242 (1961)
14. Pyke, R.: Markov renewal processes with finitely many states. Ann. Math. Stat. **32**, 1243–1259 (1961)

Chapter 15
Analysis Methods for Unreplicated Factorial Experiments

P. Angelopoulos, C. Koukouvinos, and A. Skountzou

Abstract The analysis of unreplicated designs concentrates much of interest, since these designs enable us to estimate the factorial effects using contrasts, while no degrees of freedom are left to estimate the error variance, so conventional ANOVA techniques cannot be applied to detect the active effects. In this paper we review two effective methods (Angelopoulos and Koukouvinos, J. Appl. Statist 35:277–281, 2008; Angelopoulos et al., Qual. Reliab. Eng. Int 26:223–233, 2010) for the identification of active factors in unreplicated experiments. An illustrative example of the application of the two methods is presented, as also a comparative simulation study, revealing the effectiveness of the two methods.

Keywords Two-level factorial designs • Unreplicated experiments • Outliers • Projective property • Power

AMS Subject Classification: Primary 62K15, Secondary 62J20

15.1   Introduction

Factorial designs are widely used in screening experiments where the effect of several factors on a response variable needs to be studied. A special class of these designs is 2^k factorial designs, where k factors are involved, each at only two levels. These levels can be quantitative, such as two values of temperature, high and low, or qualitative, such as two machines or two operators. In such

P. Angelopoulos • C. Koukouvinos • A. Skountzou
Department of Mathematics, National Technical University
of Athens, Zografou 15773, Athens, Greece
e-mail: pangelopoulos@math.ntua.gr; ckoukouv@math.ntua.gr; askount@central.ntua.gr

N.J. Daras (ed.), *Applications of Mathematics and Informatics in Military Science*,
Springer Optimization and Its Applications 71, DOI 10.1007/978-1-4614-4109-0_15,
© Springer Science+Business Media New York 2012

Table 15.1 A 2^3 full
factorial design with
interactions

	I	A	B	C	AB	AC	BC	ABC
(1)	+1	−1	−1	−1	+1	+1	+1	−1
a	+1	+1	−1	−1	−1	−1	+1	+1
b	+1	−1	+1	−1	−1	+1	−1	+1
ab	+1	+1	+1	−1	+1	−1	−1	−1
c	+1	−1	−1	+1	+1	−1	−1	+1
ac	+1	+1	−1	+1	−1	+1	−1	−1
bc	+1	−1	+1	+1	−1	−1	+1	−1
abc	+1	+1	+1	+1	+1	+1	+1	+1

designs we usually code the two levels as −1 and +1 and call them low and high setting. A 2^k full factorial design consists of k factors or treatments (columns) and 2^k experimental runs (rows), where each row of the design corresponds to each treatment combination. A full factorial design includes all possible combinations of the factor levels. In Table 15.1 we see the example of a 2^3 full factorial design. The first column with all elements +1 corresponds to the general mean. The next three columns correspond to the three main effects A, B, and C, and the four last columns correspond to the factors interactions, which are expressed by the product of the involved factors. As the number of experimental factors increases, the number of runs grows exponentially, and in many cases only a single replicate of the design may be allowed. A single replicate of a 2^k factorial design is usually called an unreplicated factorial design. These designs are saturated, that is, the number of the parameters to be estimated equals the number of the total runs, and no degrees of freedom are available for an independent estimation of the error variance. Unreplicated factorial designs are often employed in the initial stages of an experiment in order to identify the active factors among a large number of potential active variables and pose them in further investigation. These designs are widely used in engineering, technological, industrial, or military processes, where every experimental trial is highly expensive and only a limited number of experimental runs are available.

The analysis of unreplicated designs is a complicated problem, since these designs enable the estimation of the $2^k − 1$ factorial effects using contrasts, but no degrees of freedom are left to estimate the error variance, and conventional ANOVA techniques cannot be applied to detect the active effects. Many methods have been proposed for the analysis of unreplicated factorial designs and their performance has been evaluated and reported in the literature (see, for example, the detailed review article of Hamada and Balakrishnan [10] and the book by Voss and Wang [14]). A normal or half-normal plot of the effect estimates, as introduced by Daniel [7], is the most commonly used method in testing the effects significance. Box and Meyer [5], Lenth [11], Benski [4], Dong [8], Chen and Kunert [6], Aboukalam [1], Miller [12], Voss and Wang [15], and many other authors have proposed analysis techniques for unreplicated designs. In this context we review two methods [2, 3] for the identification of the true active effects in unreplicated factorial designs.

15.2 Two Analysis Methods for 2^k Unreplicated Designs

Consider the following model for testing the significance of $p = 2^k - 1$ effects in two-level unreplicated factorial designs:

$$\mathbf{y} = \sum_{i=0}^{p} \beta_i \mathbf{x}_i + \mathbf{e} \tag{15.1}$$

where $\mathbf{y} = (y_1, \ldots, y_n)$ is the response vector, n is the number of runs, $\mathbf{x}_i, i = 1, \ldots, p$, are the effects vectors with levels ± 1, $\mathbf{x}_0 = \mathbf{1}_n$ is the n-dimensional vector of 1's which corresponds to the general mean, $\beta_i, i = 0, \ldots, p$ are unknown parameters, and $\mathbf{e} = (e_1, \ldots, e_n)$ is a random error term. The global null hypothesis under testing is $H_0 : \beta_1 = \cdots = \beta_k = 0$.

15.2.1 A Method Based on Outliers Detection

Motivated by the subjective nature of a Daniel plot [7], the authors proposed a formal test for the identification of active effects in [2]. Consider that $\hat{\theta}_i$ are the ordered factorial effect estimates (main and interactions) and $l_i = \Phi^{-1}([i - 0.5]/p)$ their corresponding normal probability values. Then, the following data set is obtained:

X_i	l_1	l_2	$\cdots$	l_p
Y_i	$\hat{\theta}_1$	$\hat{\theta}_2$	$\cdots$	$\hat{\theta}_p$

and a linear regression model of the form:

$$Y_i = \beta_0 + \beta_1 X_i + \varepsilon_i \tag{15.2}$$

can be fitted to this data set, where β_0, β_1 are the regression coefficients and ε_i is the random error. Following Daniel's idea the active factors under model (15.2) can be detected as outliers values. The algorithm proposed by Hadi and Simonoff [9] is used for the outliers identification.

The Hadi and Simonoff Algorithm is a four steps procedure. The first step is denoted as *Step 0*, in which an initial basic subset M of no outliers is defined.

- *Step 0* Find a basic subset M with no outliers. Order the p effects according to $|a_i|$, where a_i are the adjusted residuals given by $a_i = e_i/\sqrt{1 - p_{ii}}$, where p_{ii} is the i diagonal element of the matrix $\mathbf{P} = \mathbf{X}(\mathbf{X}^T\mathbf{X})^{-1}\mathbf{X}^T$. The initial size of the set M will be 2, consisting of the effects corresponding to the smallest $|a_i|$.

- *Step 1* Compute and order the p effects according to:

$$
\begin{array}{ll}
\dfrac{|y_i - x_i^T \hat{\beta}_M|}{\sqrt{1 - x_i^T (X_M^T X_M)^{-1} x_i}} & \text{if } x_i \in M \\[4mm]
\dfrac{|y_i - x_i^T \hat{\beta}_M|}{\sqrt{1 + x_i^T (X_M^T X_M)^{-1} x_i}} & \text{if } x_i \notin M
\end{array}
\tag{15.3}
$$

Form a new basic subset that contains the first $s + 1$ effects, where s is the number of effects in the previous basic subset. Continue this process until the basic subset contains $h = [p/2]$ effects.

- *Step 2* Compute

$$
d_i = \begin{cases}
\dfrac{y_i - x_i^T \hat{\beta}_M}{\hat{\sigma}_M \sqrt{1 - x_i^T (X_M^T X_M)^{-1} x_i}} & \text{if } x_i \in M \\[4mm]
\dfrac{y_i - x_i^T \hat{\beta}_M}{\hat{\sigma}_M \sqrt{1 + x_i^T (X_M^T X_M)^{-1} x_i}} & \text{if } x_i \notin M
\end{cases}
\tag{15.4}
$$

where M is a basic subset with no active effects originally of size h.

- *Step 3* Arrange the effects in ascending order according to $|d_i|$, and let $d_{(s+1)}$ be the $(s + 1)$-ordered statistic of $|d_i|$, and s the size of the current subset M.

  1. If $d_{(s+1)} \geq t_{(a/2(s+1), s-k)}$ then declare all effects satisfying $|d_i| \geq t_{(a/2(s+1), s-k)}$ as active and stop.
  2. Otherwise, form a new basic subset M with the first $(s + 1)$-ordered effects. If $s + 1 = p$, then stop and declare no active effects; otherwise go to *Step 2*.

The observations declared as outliers according to the above procedure correspond to the active effects under model (15.1).

15.2.2 A Method Based on the Projection Property

The second method [3] is based on the projection property of factorial designs, i.e., such designs can be projected into smaller designs by the significant factors. The authors suggest determining first a set of inactive effects, in order to take advantage of the projective property, and project the factorial design in those factors that appear to be active and use the classical ANOVA techniques to perform tests. Suppose that A is the set of all factorial effects of a factorial design with k main effects, and $P_i, i = 1, \ldots, k$ are the k subsets of factorial effects obtained after projecting the unreplicated design into all possible choices of $k - 1$ factors. Each projection design can be viewed as a new experiment, which can be analyzed since there are 2^{k-1} degrees of freedom left to estimate the experimental error. The active and inactive effects in each P_i can be identified using an analysis of variance. If a factorial effect is found to be active in any projection design analysis, then it appears to be a potential

Table 15.2 Critical values

	Error rate		
p	0.01	0.05	0.1
7	0.00266	0.01112	0.0204
15	0.0009	0.0027	0.0055
31	0.00066	0.002	0.00383
63	0.000167	0.00084	0.00167
127	0.00016	0.0003	0.00071

active effect for the original unreplicated design. Similarly, if a factorial effect is found to be inactive in all k projection design analysis, then this effect is highly unlikely to be active for the unreplicated design.

Consider that AC_i, $i = 1, \ldots, k$, are the sets containing the active effects of each subset P_i. Then, the potential set AC of active effects is $AC = \bigcup_{i=1}^{k} AC_i$, while the set of inert effects is $IN = \bigcup_{i=1}^{k} P_i - AC$. The next step is to test the significance of the effects belonging to the set $A - IN$ and extract conclusions for the original unreplicated design. The authors propose that, in order to control the experimental error at a desired level, the critical value at each projection should be equal to 0.01, while the critical value for the whole experiment should be chosen according to the values in Table 15.2.

15.2.3   An Illustrative Example

Montgomery described in [13] a real experiment where the effect of four factors in the filtration rate of a chemical process is studied. A 2^4 full factorial design in a single replicate is used to conduct the experiment and data are presented in Table 15.3. Various analysis techniques have identified the effects A, C, D, AC, AD being the only active effects in the experiment. The application of the two methods described above results to the same conclusions.

First, we analyze the example using the method based on outliers presented in Sect. 15.2.1. We calculate the estimates of the factorial effects and their corresponding normal probability values, as in Table 15.4. We apply a simple linear regression model to the data set of Table 15.4 and form the basic subset corresponding to the effects with the two smallest adjusted residual values, which are B and ABD. We apply the first step of the method until the basic subset has seven effects. At the end of the first step, AB, ABCD, BD, ABC, CD, BC, ACD are added to the basic subset. Proceeding to the second step, we calculate the d_i statistics. In step 3 we compare the $s + 1$ d_i statistic with the value of the t-distribution. For the first iteration of the method we compare the eighth-ordered statistic. This process is repeated until a d_i, such that $|d_i|$ is greater than the value of the t-distribution, is found. This is done after the fourth iteration and the effects A, C, D, AC, AD are declared active.

Table 15.3 Data of the chemical process example

Run	A	B	C	D	y
1	−	−	−	−	45
2	+	−	−	−	71
3	−	+	−	−	48
4	+	+	−	−	65
5	−	−	+	−	68
6	+	−	+	−	60
7	−	+	+	−	80
8	+	+	+	−	65
9	−	−	−	+	43
10	+	−	−	+	100
11	−	+	−	+	45
12	+	+	−	+	104
13	−	−	+	+	75
14	+	−	+	+	86
15	−	+	+	+	70
16	+	+	+	+	96

Table 15.4 Estimates and normal probability values

Effect	Estimate	$\Phi^{-1}([i-0.5])/p$	Effect	Estimate	$\Phi^{-1}([i-0.5])/p$
AC	−18.125	−1.83391	BC	2.375	0.167894
BCD	−2.625	−1.28155	B	3.125	0.340695
ACD	−1.625	−0.967422	ABD	4.125	0.524401
CD	−1.125	−0.727913	C	9.875	0.727913
BD	−0.375	−0.524401	D	14.625	0.967422
AB	0.125	−0.340695	AD	16.625	1.28155
ABCD	1.375	−0.167894	A	21.625	1.8339
ABC	1.875	0			

Next, we analyze the data from Table 15.3 using the method based on the projection property of the factorial designs (see Sect. 15.2.2). The four projections of the 2^4 full factorial design are $P_1 = \{B,C,D,BC,BD,CD,BCD\}$, $P_2 = \{A,C,D,AC,AD,CD,ACD\}$, $P_3 = \{A,B,D,AB,AD,BD,ABD\}$, $P_4 = \{A,B,C,AB,AC,BC,ABC\}$. Each projection consists of seven effects and the general mean and eight degrees of freedom are available for the estimation of the error variance. We apply regular ANOVA to each projection by setting the critical value equal to 0.01 as suggested by the authors. The active set AC_i of every subset P_i is defined by comparing the corresponding p-values with 0.01. The resulted p-values in each projection are reported in Table 15.5. The sets of active effects are $AC_1 = \{\}$, $AC_2 = \{A,C,D,AC,AD\}$, $AC_3 = \{A\}$, $AC_4 = \{A\}$.

Table 15.5 P-values of the effects in each of the four projection designs

Effect	P1	P2	P3	P4
A	–	0	0.00202	0.0027
B	0.79	–	0.6872	0.708
AB	–	–	0.9871	0.988
C	0.42	0.003		0.2555
AC	–	0.0001		0.0547
BC	0.844	–		0.7758
ABC	–	–		0.822
D	0.247	0.0003	0.0864	–
AD	–	0.0001	0.057	–
BD	0.975	–	0.9613	–
ABD	–	–	0.5965	–
CD	0.925	0.647	–	–
ACD	–	0.512	–	–
BCD	0.828	–	–	–
ABCD	–	–	–	–

So, the set of possible active effects is $AC = \{A,C,D,AC,AD\}$, while the set of inert effects is $IN = \{B,AB,BC,BD,CD,ABC,ABD,ACD,BCD,ABCD\}$. Proceeding in an ANOVA, excluding the effects in the set IN, results in the identification of A, C, D, AC, AD (have p-values less than 0.0027) as active effects.

15.3 Evaluation of the Methods

We use the notion of Power in three forms for the comparison of the two methods and Lenth's test [11]. Power is defined as the expected fraction of active effects that are declared active and has been used by many authors (see [6, 10]) as a measure of performance. Power I expresses the probability of rejecting the null hypothesis that all the factorial effects are inactive. Power II is defined as the probability that the effects declared being active include all really active contrasts.

Ten thousand simulated experiments are conducted for each case, where a response vector y is generated according to model (15.1) for each experiment, the 2^4 unreplicated factorial design serves as the design matrix, the errors are i.i.d.'s with $N(0,\sigma)$ and σ is set equal to 1. We consider the cases for $p = 1$ up to seven active factorial effects involving in the experiments having the same magnitude equal to 2σ and the results are presented in Figs. 15.1–15.3. It is shown that both methods outperform Lenth's method with respect to the values of Power, and the second method achieves the higher values among the three methods. The two methods denote high values of Power I even in the cases of 6 and 7 factors. They also appear very powerful in Fig. 15.3, where clearly they have better values than Lenth's method.

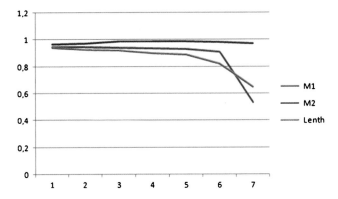

Fig. 15.1 Power of the two methods compared to Lenth's test

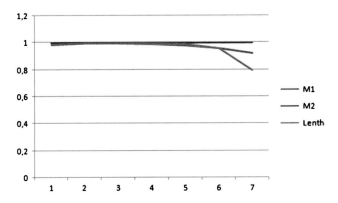

Fig. 15.2 Power I of the two methods compared to Lenth's test

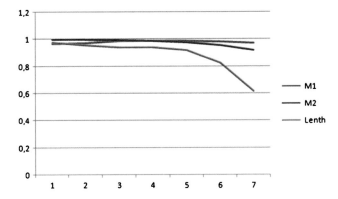

Fig. 15.3 Power II of the two methods compared to Lenth's test

References

1. Aboukalam, M.A.F.: Quick, easy and powerful analysis of unreplicated factorial designs. Commun.Statist.- Theory Meth. **34**, 1169–1175 (2005)
2. Angelopoulos, P., Koukouvinos, C.: Detecting active effects in unreplicated designs, J. Appl. Statist. **35**, 277–281 (2008)
3. Angelopoulos, P., Evangelaras, H., Koukouvinos, C.: Analyzing unreplicated 2^k factorial designs by examining their projections into $k - 1$ factors. Qual. Reliab. Eng. Int. **26**, 223–233 (2010)
4. Benski, H.C.: Use of a normality test to identify significant effects in factorial designs. J. Qual. Tech. **21**, 174–178 (1989)
5. Box, G.E.P., Meyer, R.D.: An analysis for unreplicated fractional factorials. Technometrics. **28**, 11–18 (1986)
6. Chen, Y., Kunert, J.: A new quantitative method for analysing unreplicated factorial designs. Biometrical J. **46**, 125–140 (2004)
7. Daniel, C.: Use of half-normal plots in interpreting factorial two-level experiments. Technometrics. **1**, 311–341 (1959)
8. Dong, F.: On the identification of active contrasts in unreplicated fractional factorials. Stat. Sinica. **3**, 209–217 (1993)
9. Hadi, A.S., Simonoff, J.S.: Procedures for identification of multiple outliers in linear models. J. Am. Stat. Association. **88**, 1264–1271 (1993)
10. Hamada, M., Balakrishnan, N.: Analysing unreplicated factorial experiments: A review with some new proposals. Stat. Sinica. **8**, 1–41 (1998)
11. Lenth, R.V.: Quick and easy analysis of unreplicated factorial. Technometrics. **31**, 469–473 (1989)
12. Miller, A.: The analysis of unreplicated factorial experiments using all possible comparisons. Technometrics. **47**, 51–63 (2005)
13. Montgomery, D.C.: Design and analysis of experiments, 6th edn. John Willey and Sons, New York (2009)
14. Voss, D.T., Wang, W.: Analysis of othogonal saturated designs. In: Screening Methods for Experimentation in Industry, Drug Discovery and Genetics, pp. 268–281. Springer, New York (2006)
15. Voss, D.T., Wang, W.: On adaptive testing in orthogonal satutated designs. Stat. Sinica. **16**, 227–234 (2006)

Printed by Publishers' Graphics LLC
SO20120821

.